# Montmorillonite Clay

*Edited by Faheem Uddin*

Published in London, United Kingdom

IntechOpen

*Supporting open minds since 2005*

Montmorillonite Clay
http://dx.doi.org/10.5772/intechopen.92926
Edited by Faheem Uddin

Contributors
Hakan Ciftci, Chakradhar V.P. Vp Komanduri, Reda Marouf, Nacer Dali, Nadia Boudouara, Fatima Ouadjenia, Faiza Zahaf, Boukhlifi Fatima, Kali Abderrahim, Loulidi Ilyasse, Amar Abdelouahed, Hadey Chaimaa, Jabri Maria, Mbarka Ouchabi, Faheem Uddin



First published in London, United Kingdom, 2021 by IntechOpen
IntechOpen is the global imprint of INTECHOPEN LIMITED, registered in England and Wales, registration number: 11086078, 5 Princes Gate Court, London, SW7 2QJ, United Kingdom
Printed in Croatia

British Library Cataloguing-in-Publication Data
A catalogue record for this book is available from the British Library

Additional hard and PDF copies can be obtained from orders@intechopen.com

Montmorillonite Clay
Edited by Faheem Uddin
p. cm.
Print ISBN 978-1-83968-602-3
Online ISBN 978-1-83968-603-0
eBook (PDF) ISBN 978-1-83968-604-7

# Meet the editor

Faheem Uddin earned a Ph.D. in Textile Special Finishing from the University of Manchester, UK. His current research interests are in the processing of the textile fiber, clay study, flame retardancy, cellulose/cotton cross-linking, and environment-friendly processing. He is the principal author of more than 40 research publications in reputed journals. He is the author of the Higher Education Commission (HEC) award-winning research paper on montmorillonite clay that is a highly cited paper to date and has over 26 000 ResearchGate reads. He is a fellow of The Textile Institute, United Kingdom, and received the Best Research Paper Award by HEC, Pakistan, and the Research Productivity Award, PSF, Pakistan, and RULA Best Researcher of The Year 2019. He is presently serving as a professor at Iqra University, Karachi, Pakistan.

# Contents

**Preface** XI

**Section 1**
Introduction 1

**Chapter 1** 3
Introductory Chapter: Montmorillonite Clay Consumption Trend in Industry
*by Faheem Uddin*

**Section 2**
Structure and Properties 11

**Chapter 2** 13
Study of Adsorption Properties of Bentonite Clay
*by Reda Marouf, Nacer Dali, Nadia Boudouara, Fatima Ouadjenia and Faiza Zahaf*

**Section 3**
Montmorillonite Composite 31

**Chapter 3** 33
Reinforcement of Montmorillonite Clay in Epoxy/Unsaturated Polyester Blended Composite: Effect on Composite Properties
*by Chakradhar V.P. Komanduri*

**Section 4**
Industrial Application 41

**Chapter 4** 43
Exploitation of Bentonite for Wastewater Treatment
*by Kali Abderrahim, Loulidi Ilyasse, Amar Abdelouahed, Boukhlifi Fatima, Hadey Chaimaa, Jabri Maria and Mbarka Ouchabi*

**Section 5**
Testing and Evaluation 63

**Chapter 5** 65
An Introduction to Montmorillonite Purification
*by Hakan Ciftci*

# Preface

In an ever-expanding world of innovation and technology, the significance of environmental sustainability is increasingly important for life and survival of mankind on planet Earth. The material and structure of a product and the processes it undergoes are now checked to know whether the effects are supportive to our natural environment and its sustainability. It would be unwise to compromise the good effects. The materials that can offer the desired performance without harming the natural environment are obviously viable for the sustainable growth of the industry. Consequently, montmorillonite (MMT) clay becomes an attractive material in making various products and composite structures.

The book chapters are written to provide scientific understanding and knowledge to scholars, students, faculty members, and laboratory and industry workers of the subject of montmorillonite clay, including its structure and properties, use in composite, industrial application, and testing and evaluation. The introduction provides a useful discussion on the current consumption trend in the industrial market. This is to signify how important montmorillonite clay is as a material in the modern and emerging market of product development and industrial applications.

In the chapter on the purification of montmorillonite, the process of removing non-clay minerals (gangue), such as calcite, feldspar, quartz, opal-CT, and mica, from montmorillonite ore is introduced. Physical and chemical purification processes are discussed. Such purification is important for applications in pharmaceutical, cosmetic, food, and advanced materials for nanotechnology. The chapter on using bentonite in wastewater treatment presented the efficiency of bentonite and modified bentonite to purify aqueous solutions containing organic pollutants, such as phenol.

Montmorillonite composite is presently an important subject that is covered in the chapter *Reinforcement of Montmorillonite Clay in Epoxy/Unsaturated Polyester Blended Composite: Effect on Composite Properties*. This chapter discusses the montmorillonite clay dissemination into unsaturated polyester (UP) and epoxy blend systems in diverse weight ratios to prepare epoxy/UP/ MMT clay composite. The prepared specimens were characterized by thermal and chemical analysis.

An analytical study of using activated bentonite is presented in the chapter *Study of Adsorption Properties of Bentonite Clay*. This is an important chapter to learn the physicochemical properties of bentonite in adsorption using analytical techniques, such as chemical composition, X-ray diffraction (XRD), Fourier transform infrared spectroscopy (FTIR), and Brunauer–Emmett–Teller surface area (SBET). The bentonite was intercalated by aluminum poly-cation solution and cetyltrimethylammonium bromide. The acid activation of natural bentonite was done by treatment with hydrochloric acid. The surface water pollutants removed by the prepared bentonite were BEMACID YELLOW E-4G and reactive MX-4R dye and fungicide chlorothalonil.

Optimistically, concerned readers will find the content of the book useful in knowing and understanding montmorillonite clay.

Finally, I am grateful to all the authors, commissioning editor, and the supporting staff at the IntechOpen team for their valuable contribution and for making this book project successful.


**Dr. Faheem Uddin**
Professor,
Asian Institute of Fashion Design (AIFD),
Iqra University,
Karachi, Pakistan

Section 1

# Introduction

Chapter 1

# Introductory Chapter: Montmorillonite Clay Consumption Trend in Industry

*Faheem Uddin*

## 1. Introduction

Clay is the material of choice in the 21st century. The value of natural environment and sustainability is now recognized as more demanding at social, political, and public levels. The realization of these two prime influencers has resulted in an increased demand for montmorillonite consumption. The performance features in montmorillonite clay in meeting the application requirement for a variety of materials is the foundation for the continual interest in production, consumption, and research studies.

Clay utilization from grave flooring to pottery and construction of bricks, with origin traced several thousand years back, is breathing to date in human society. Increasing interest and realization in saving the natural environment, through using environment-friendly material, will apparently capture the motivation and attention of researchers and industrialists to value the clay consumption in products and structure of materials.

The earliest examples of clay consumption as a product or constituent material can be seen in the oldest human civilization. The abundant presence of clay in the natural environment is a vital source of directing the interest in using the clay as a product and as an effective material component in a composite structure.

Presently, the study of montmorillonite clay structure and properties [1], industrial applications, composite materials, and testing and evaluation are active areas for research and development studies.

More importantly, the sustainability, environment-friendly character, cost-effectiveness, and ease of processing are continually supporting an enhanced interest in consuming clay for the development of a variety of products (**Figure 1**).

The study can be seen presenting an overview of montmorillonite clay structure and properties, and addressing its effects in selected materials [2]. Important aspects of montmorillonite clay utilization for the industry are discussed in terms of natural sources, chemical structure, and physical and chemical properties. The useful properties for industrial application include, however, not limited to, particle size and layered structure, molecular structure and cation exchange effect, barrier property, and water sorption.

It is fascinating to see clay as the oldest material consumed in human civilization, and in the 21st century, it appears to be the latest material in modern human society.

Clay as a material has an extensive variety of uses and application [3–6]. Many of the uses of clay are traditional; however, an increasing interest in clay research and innovation is providing continuous breathing to its global market consumption.

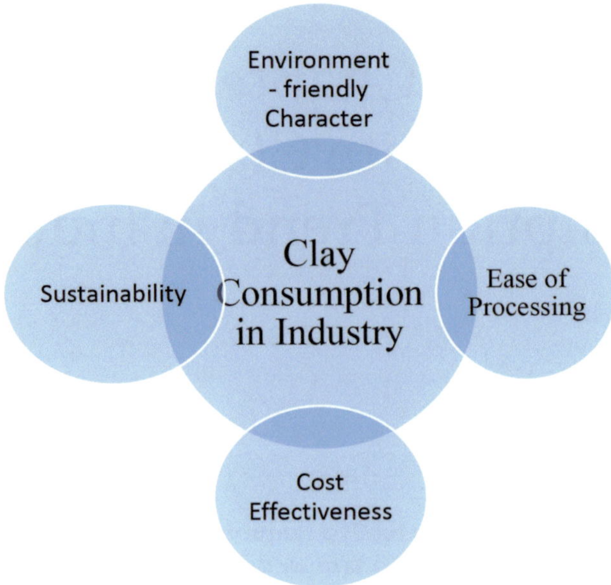


**Figure 1.**
*Important driving factors leading the clay consumption in the industry.*

The interesting clay consumption in traditional sectors including civil structure, paint, construction products, plastic, etc., coupled with more specialized uses in medical, healthcare, and industrial applications demonstrate the current and future viability of clay consumption.

The study of clay consumption in the research studies can be seen for particular application materials and products. However, any quantified information available on clay consumption in the industrial market is mainly seen in the progress studies of industrial products. This chapter provides an overview of the montmorillonite clay consumption mainly based on the reports produced by the market studies, where necessary reference is made to the research studies that direct the particular use or consumption of montmorillonite.

## 2. Montmorillonite clay producers

The consumption of clay in the industry is presented in reports by various study groups on a commercial basis. These reports demonstrate the clay consumption in various industrial sectors in terms of quantity and value. Montmorillonite clay is generally represented by sodium bentonite and calcium bentonite in the industrial sector.

Undoubtedly, commercially available industrial clay is a chemical composition that generally comprises minerals, metal oxide, and organic traces [7]. The industrial market studies are showing the growth in montmorillonite clay consumption over the period of 2012–2030. The main known producers of montmorillonite clay around the world are apparently the same; they can be seen in **Table 1**. The influence of the COVID-19 pandemic has not affected the presence of producers; however, depending upon the country or region, the variation in production is obviously possible.

It can be seen that for the last almost two decades, the main montmorillonite clay producers for the industry are there in different study reports. These producers are mainly located in the USA, China, India, and Turkey (countries with more than

| Montmorillonite clay producers | Projected year | Reference |
|---|---|---|
| Amcol (Minerals Technologies); Bentonite Performance Minerals LLC (US); Wyo-Ben Inc. (US); Black Hills Bentonite (US); Tolsa Group (Spain); Imerys (S&B) (France); Clariant (Switzerland); Bentonite Company LLC (Russia); Laviosa Minerals (Italy); LKAB Minerals (Netherlands); Ashapura (India); Star Bentonite Group (India); Kunimine Industries (South America); Huawei Bentonite (China); Fenghong New Material (China); Chang'anRenheng (China); LiufangziBentonite (China); BentonitUniao (Brazil); Castiglioni (Argentina), Canbensan (Turkey) | 2021–2030 | https://www.marketwatch.com/press-release/montmorillonite-clay-market-consumption-companies-and-industry-report-2021-2030-marketbiz-2021-06-18 [8] |
| Amcol (Minerals Technologies); Bentonite Performance Minerals LLC(US); Wyo-Ben Inc. (US); Black Hills Bentonite (US); Tolsa Group (Spain); Imerys (SandB) (France); Clariant (Switzerland); Bentonite Company LLC (Russia); Laviosa Minerals (Italy); LKAB Minerals (Netherlands); Ashapura (India); Star Bentonite Group (India); Kunimine Industries (South America); Huawei Bentonite (China); Fenghong New Material (China); Chang'anRenheng (China); LiufangziBentonite (China); BentonitUniao (Brazil); Castiglioni (Argentina); Canbensan (Turkey); Aydin Bentonit (Turkey); KarBen (Turkey); G and W Mineral Resources (South Africa); NingchengTianyu (China) | 2021–2026 | https://www.wboc.com/story/43563723/montmorillonite-clay-market-2021-is-estimated-to-clock-a-modest-cagr-of-40nbspduring-the-forecast-period-2021-2026-with-top-countries-data [9] |
| Amcol (Minerals Technologies); Bentonite Performance Minerals LLC (US); Wyo-Ben Inc. (US); Black Hills Bentonite (US); Tolsa Group (Spain); Imerys (S&B) (France); Clariant (Switzerland); Bentonite Company LLC (Russia); Laviosa Minerals SpA (Italy); LKAB Minerals (Netherlands); Ashapura (India); Star Bentonite Group (India); Kunimine Industries (South America); Huawei Bentonite (China); Fenghong New Material (China); Changan Renheng (China); Liufangzi Bentonite (China); Bentonit Uniao (Brazil); Castiglioni Pesy Cia (Argentina); Canbensan (Turkey); Aydn Bentonit (Turkey); KarBen (Turkey); G & W Mineral Resources (South Africa); NingchengTianyu (China) | 2020–2027 | https://www.marketresearchintellect.com/product/global-montmorillonite-clay-bentonites-consumption-market-size-and-forecast/ [10] |
| Amcol (Minerals Technologies); Bentonite Performance Minerals LLC (US); Wyo-Ben Inc. (US); Black Hills Bentonite (US); Tolsa Group (Spain); Imerys (S&B) (France); Clariant (Switzerland); Bentonite Company LLC (Russia); Laviosa Minerals SpA (Italy); LKAB Minerals (Netherlands); Ashapura (India); Star Bentonite Group (India); Kunimine Industries (South America); Huawei Bentonite (China); Fenghong New Material (China); Changan Renheng (China); LiufangziBentonite (China); Bentonit Uniao (Brazil); Castiglioni Pesy Cia (Argentina); Canbensan (Turkey); AydinBentonit (Turkey); KarBen (Turkey); G & W Mineral Resources (South Africa); NingchengTianyu (China) | 2012–2022 | https://www.marketresearch.com/QYResearch-Group-v3531/Global-Montmorillonite-Clay-Bentonites-Research-11763149/ [11] |

**Table 1.**
*The main producers of montmorillonite clay around the world over the period 2012–2030.*

one producing company). Countries severely affected by the COVID-19 pandemic including Italy, India, Spain, and the USA may be having some variation in the production capacity. The main influence of the COVID-19 pandemic is possible on the demand, production, and supply chain of montmorillonite clay.

## 3. Variety of industrial clay

Naturally occurring bentonite is essentially comprising montmorillonite. Bentonite contains exchangeable cations including $Ca^{2+}$, $Mg^{2+}$, $Na^{+}$, or $Li^{+}$. The dominant presence of a particular cation in bentonite provided important application properties. Sodium bentonite can swell in water. It has good binding property and is generally used in its natural state. However, calcium bentonite has a non-swelling character. The binding property in calcium bentonite is achievable through a chemical reaction with soda ash to introduce sodium-exchanged bentonite at outer and inner surfaces, resulting in an increased ability to bind water. Calcium bentonite can also be treated with acid to obtain bleaching clay. Li bentonite is called hectorite; however, it is not mined at a commercial level like sodium or calcium bentonite [12].

## 4. Montmorillonite clay applications

The variety of montmorillonite clay applications is significantly diverse. It is applied in areas of catalysis, wastewater treatment, food additive, antibacterial function, polymer, sorbent, etc. Some more specific applications include pelletizing (production of iron-ore pellets, civil engineering and construction work, drilling, impurities removal in oils, animal feed, producing paste composition in pharmaceutical, cosmetics and medical products, detergents, papers, etc. [13].

The significant development in the use and application of montmorillonite is seen in recent times. This chapter provides an overview of montmorillonite's structure and properties and particularly discusses its recent utilization in important materials. Montmorillonite is introduced in terms of its natural sources, chemical structure, physical and chemical properties, and functional utilization. The important physical and chemical properties are summarized as particle and layered structure, molecular structure and cation exchange effect, barrier property, and water sorption. This is followed by the important functional utilizations of montmorillonite based on the effects of its chemical structure. The important functional utilization of montmorillonite includes use as food additive for health and stamina, for antibacterial activity against tooth and gum decay, as sorbent for nonionic, anionic, and cationic dyes, and the use as a catalyst in organic synthesis. In terms of healthcare, the effects of montmorillonite clay are observed for body detoxification, resisting skin allergy and dermatitis, treatment of organism leading to diarrhea, antibacterial effects (killing large spectrum of bacteria), assisting renal health, providing drug delivery for cancer treatment, etc. [14].

The environmental concerns, to date, do not indicate the adversity for particles used as additives. Studies will be useful that are clearly based on any montmorillonite structure to describe environmental effects.

The broader application categories described in the market studies include molding sands, iron-ore pelletizing, pet litter, drilling mud, civil engineering, and agriculture.

## 5. Montmorillonite market

For the last two decades, the trend in the industrial consumption of montmorillonite clay is progressive, and there are reports indicating the continuity of this trend for the next decade.

Montmorillonite clay market is divided into the United States, Mexico, Canada, Germany, Singapore, the U.K., Italy, Russia, France, Spain, China, India, Japan, South

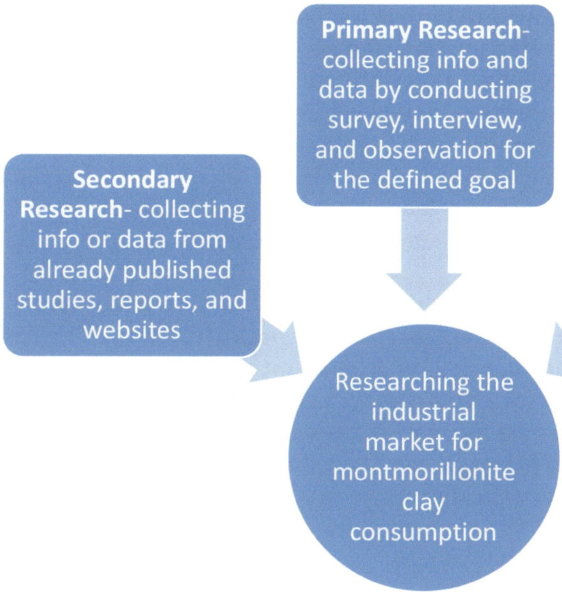


**Figure 2.**
*Sources for the study of industrial market study.*

Korea, Australia, Brazil, Colombia, Paraguay, Saudi Arabia, South Africa, Egypt, the UAE, and ASEAN countries. The details info and data can be seen in the market study reports, covering the worldwide montmorillonite clay (bentonites) market, the competitiveness of suppliers, market share, size, development rate, future patterns, etc. The discussion for future market trends and the factors that are driving the montmorillonite clay market are provided [15].

The study of the industrial market is based on research methodology that covers three important sources including primary research, secondary research, and expert panel reviews. Where secondary research utilized the sources including press releases, company annual reports, and research papers of the concerned industry. The study is refined using the associated sources of industry magazines, trade journals, websites of related government departments, and the trade associations [10].

The worldwide montmorillonite clay market, for period of 2020–2027, is estimated to move from USD 1238.7 million (2020) to USD 1630 million by 2027; the growth is forecasted at a CAGR of 4.0% (**Figure 2**) [16].

Reducing consumption trend in montmorillonite clay was seen in 2008–2009. However, start recovering the market in 2010. For example, global production of bentonite was reduced by a further 5% in 2010 relative to 2009, and a reduction of 9% in 2009 relative to 2008. The fall was perceived as the decreased demand by the end-user market.

In the following year, the consumption was recovering. Therefore, bentonite production of more than 11 million tons was globally observed in 2011. US Geological Survey estimated the worldwide reserves to be over 10 billion tons [17].

## 6. Conclusion

Montmorillonite clay is an important material to produce a large variety of products for the industrial consumption around the world. In addition to meeting the requirement for the environment and sustainability, the montmorillonite clay is known for its inexpensive, abundance, and natural character. It is fascinating to see clay as the oldest material consumed in human civilization, and in the 21st century, it appears to be the latest material in modern human society.

Clay as a material has an extensive variety of uses and application. Many of the uses of clay are traditional; however, an increasing interest in clay research and innovation is providing continuous breathing to its global market consumption. The interesting clay consumption in traditional sectors including civil structure, paint, construction products, plastic, etc., coupled with more specialized uses in medical, healthcare, and industrial applications demonstrate the current and future viability of clay consumption.

The reported research literature introduced the use of montmorillonite clay in particular application, or demonstrate its effect in specific material. The data on the consumption of montmorillonite clay is based on products categorized by the industrial application. Therefore, this chapter provides an overview of montmorillonite clay consumption using the market research reports. It indicates the important market regions, suppliers, industrial product areas and applications of montmorillonite clay. If necessary the discussion is supported with the particular research study.

## Conflict of interest

There is no conflict of interest in the publication of this manuscript.

## Author details

Faheem Uddin
Asian Institute of Fashion Design, Iqra University, Karachi, Pakistan

*Address all correspondence to: dfudfuca@yahoo.ca

## References

[1] Uddin F. Clay, nanoclay and montmorillonite minerals. Metallurgical and Materials Transactions A. 2008;**39**(12):2804-2814

[2] Uddin F. Montmorillonite: An introduction to properties and utilization. In: Zoveidavianpoor M, editor. Current Topics in the Utilization of Clay in Industrial and Medical Applications. Mansoor Zoveidavianpoor, ed.; London, UK: IntechOpen; 2018. DOI: 10.5772/intechopen.77987

[3] Agarwal A, Raheja A, Natarajan TS, Chandra TS. Effect of electrospun montmorillonite-nylon 6 nanofibrous membrane coated packaging on potato chips and bread. Innovative Food Science and Emerging Technologies. 2014;**26**:424-430

[4] Robinson A, Johnson NM, Strey A, Taylor JF, Marroquin-Cardona A, Mitchell NJ. Calcium montmorillonite clay reduces urinary biomarkers of fumonisin $B_1$ exposure in rats and humans. Food Additives and Contaminants: Part A. 2012;**29**(5):809-818. DOI: 10.1080/19440049.2011.651628

[5] Maria HJ, Lyczko N, Nzihou A, Joseph K, Mathew C, Thomas S. Stress relaxation behavior of organically modified montmorillonite filled natural rubber/nitrile rubber nanocomposites. Applied Clay Science. 2014;**87**:120-128. DOI: 10.1016/j.clay.2013.10.019

[6] Ma J, Lei Y, Khan MA, Wang F, Chu Y, Lei W, et al. Adsorption properties, kinetics & thermodynamics of tetracycline on carboxymethyl-chitosan reformed montmorillonite. International Journal of Biological Macromolecules. 2019;**124**:557-567

[7] Anon. 2021. Available from: https://www.transparencymarketresearch.com/industrial-clay-market.html [Accessed: 2 September 2021]

[8] Anon. 2021. Available from: https://www.marketwatch.com/press-release/montmorillonite-clay-market-consumption-companies-and-industry-report-2021-2030-marketbiz-2021-06-18 [Accessed: 2 September 2021]

[9] Anon. 2021. Available from: https://www.wboc.com/story/43563723/montmorillonite-clay-market-2021-is-estimated-to-clock-a-modest-cagr-of-40nbspduring-the-forecast-period-2021-2026-with-top-countries-data [Accessed: 2 September 2021]

[10] Anon. 2021. Available from: https://www.marketresearchintellect.com/product/global-montmorillonite-clay-bentonites-consumption-market-size-and-forecast/ [Accessed: 2 September 2021]

[11] Anon. 2021. Available from: https://www.marketresearch.com/QYResearch-Group-v3531/Global-Montmorillonite-Clay-Bentonites-Research-11763149/ [Accessed: 2 September 2021]

[12] Anon. 2021. Available from: https://www.absolutereports.com/enquiry/request-sample/13870613 [Accessed: 2 September 2021]

[13] EUBA (European Bentonite Association). What is Bentonite. Available from: www.ima-europe.eu [Accessed: 2 September 2021]

[14] Moosavi M. Bentonite clay as a natural remedy: A brief review. Iranian Journal of Public Health. 2017;**46**(9):1176-1183

[15] Anon. 2021. Available from: https://www.ktvn.com/story/44226362/Montmorillonite-Clay-(Bentonites)-Market-Size-2021-with-Top-Countries-Analysis,-COVID-19-Impact-on-Players,-Industry-by-Share,-Global-Trends,-Demand-and-Future-Scope-Forecast-to-2025 [Accessed: 2 September 2021]

[16] Anon. 2021. Available from: https://www.americanrodeo.com/story/44616387/global-montmorillonite-clay-market-to-reach-usd-1630-million-growing-at-cagr-of-4-forecast-period-2021-2027 [Accessed: 2 September 2021]

[17] Hughes E. Bentonite: A swelling market. 2012. Available from: https://www.indmin.com/Article/3122191/Bentonite-a-swelling-market.html [Accessed: 2 September 2021]

## Section 2

# Structure and Properties

Chapter 2

# Study of Adsorption Properties of Bentonite Clay

*Reda Marouf, Nacer Dali, Nadia Boudouara, Fatima Ouadjenia and Faiza Zahaf*

## Abstract

The clay used in this study was the bentonite from Mostagnem, Algeria. This material is used in many fields such as drilling, foundry, painting, ceramics, etc. It can also be applied in the treatment of wastewaters from chemical industries by means of adsorption. In this chapter the physicochemical properties of bentonite were determined by using several analyses techniques such as chemical composition, XRD, FTIR and $S_{BET}$. The bentonite was intercalated by aluminum polycations solution and cethytrimethyl ammonium bromide. The acid activation of natural bentonite was performed by treatment with hydrochloric acid at different concentrations. The surface water pollutants removed by the modified bentonites are bemacid yellow E-4G and reactive MX-4R dyes, and fungicide chlorothalinil. The Langmuir and Freundlich adsorption models were applied to describe the related isotherms. The pseudo-first order and pseudo-second order kinetic models were used to describe the kinetic data. The changes of enthalpy, entropy and Gibbs free energy of adsorption process were also calculated.

**Keywords:** montmorillonite, bentonite, adsorption, dyes, fungicide

## 1. Introduction

Bentonite clay is a volcanic rock that was deposited as volcanic ash in fresh or salt water millions of years ago. It was first discovered in 1890 in Wyoming (Montana, USA) near the Fort Benton site, hence its name. Currently there are a large number of bentonite deposits in the world, in USA, in Europe, in North Africa, in Japan, in China ... In Algeria, the most economically important bentonite deposits are found in the west. We note in particular the quarry of Maghnia (Tlemcen) with reserves estimated at one million tons and that of M'zila (Mostaganem) with reserves of two million tons [1, 2].

Bentonite is very soft plastic clay composed mainly of montmorillonite, a clay mineral 2: 1 type of phyllosilicate family formed of fine particles. Montmorilonite consists of two tetrahedral layers ($SiO_4$) separated by an octahedral layer ($Al(OH)_6$), its chemical formula is $(Al, M^{2+})_2\ Si_4\ O_{10}\ (OH)_2$, n $H_2O$ with $M^{2+}$ = Mg, Fe, ... (**Figure 1**) [3]. The total thickness of the sheet is approximately 14 Å. Montmorillonites has a negative charge on the surface, neutralized by compensating cations, the main origin of this surface charge comes from isomorphic substitutions resulting from the replacement of the metal cations of the lattice by cations of the same size but of lower charge (the substitution $Al^{3+}$ by $Mg^{2+}$ or $Fe^{2+}$ and the substitution of $Si^{4+}$ by $Al^{3+}$).

 

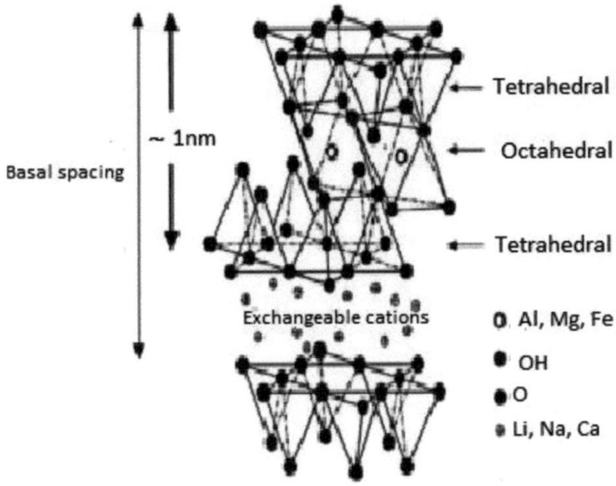


**Figure 1.**
*Montmorillonite structure.*

Bentonite has a privileged place in the purification of water [4, 5]. High specific surface area, chemical and mechanical stabilities, layered structure, high cation exchange capacity (CEC), tendency to hold water in the interlayer sites, and the presence of Brønsted and Lewis acidity have made clays excellent adsorbent materials [6, 7]. The chemical nature and pore structure of bentonite generally determine their adsorption ability [8, 9].

The adsorption properties of natural clay minerals can often be improved either by intercalation of organic, inorganic, or organometallic molecules in the interlamellar space, or by heat or acid treatments [10, 11]. XRD measurements show that intercalation increases the spacing between layers. Among these modifications, the insertion of clays by poly cationic solutions has been used widely in recent years [12, 13]. In addition to their catalytic power, they are presented as excellent adsorbents. The main qualities that these pillared clays can develop are a high specific surface area and a large pore volume.

For the poly cation intercalation pillars to execute correctly, the synthesis protocol must pass through three steps: the first is to prepare the pillaring solution based on multivalent cation such as $Al^{3+}$, $Fe^{3+}$ or $Cr^{3+}$. The second is the impregnation of the polycationic solution within the interlayer space [10, 14]. The last step is the calcination at temperature between 400 and 500 °C in order to solidify the pillars [15]. Some examples for montmorillonite intercalating by polycations cited in literature are the following: china montmorillonite was pillared by solution of $Al_{13}$ polycations to eliminate cadmium [16], the bentonite of Turkish origin pillared with Fe and Cr was applied as adsorbent for carbon oxide ($CO_2$) and hydrogen $H_2$ [17], the Al/Ti and Al/Zr- pillared montmorillonites from Brazilian Amazon was used for zinc cation removal [18] and Fe/Zr-pîllared montmorilointe (Guangdong, China) was tested for Cr(VI) removal [19].

The organic intercalation of montmorillonite can significantly enhance the organophilicity of the resultant product. Cationic organic compounds, such as surfactant cations, exchange the interlayer cations of montmorillonite [20], and the resulting organoclay is excellent for diverse organic pollutants, e.g. phenol [21], dye [22], and VOCs [23, 24]. Quaternary alkylammonium compounds (bromides or chlorides) are the most commonly employed for intercalation into bentonites. Due to their long-chain they offered to the clays a larger interlayer spacing. The most quaternary ammonium surfactants used are cethyltrimethyl ammonium bromide

(CTMAB), tetramethyl ammonium (TMA), trimethyl-phenyl ammonium (TMPA), dodecyl trimethyl ammonium and dimethyl sulfoxide (DMSO).

Unfortunately, sometimes the insertion of a surfactant weakens the specific surface of an organo-clay. To remedy this disadvantage, it's preferable to combine the intercalation of organic pillar with another mineral. For example, Jiang et al. used hexadecyltrimethyl ammonium bromide added to Al/Fe-pillared montmorillonite [25]. Zhu's group prepared an intercalated montmorillonite with both hexadecyltrimethyl ammonium bromide and Al13 cations, and discussed the influence of the addition sequence of these two modifiers on the final products [26].

Another method applied to improve adsorption capacity of bentonite consists of the reaction of clay minerals with a strong mineral acid solution, usually hydrochloric acid (HCl) or sulfuric acid ($H_2SO_4$). Acid activation leads to modification of Mt. with improved properties such as enhanced surface area, pore diameters, number of acid sites, and catalytic activities. The treatment of the naturally occurring and purified Mt. with hot mineral acid replaces exchangeable cations with $H^+$ ions. Gradual leaching of $Al^{3+}$ out of both tetrahedral and octahedral sites occurs, but the silicate groups remain largely intact [27, 28].

Dyes released into wastewaters of different industrial plants such as dye manufacturing, textile, paper and food, are toxic, not degradable and stable. The presence of these substances in the surface waters can be carcinogenic and causes damage to human beings, such as dysfunction of kidneys and central nervous system [29, 30].

Pesticides are synthesized substances or biological agents used for attracting any pest. They are mainly applied in agriculture to protect crops from insects, weeds, and bacterial or fungal diseases during growth [31]. Some pesticides, like fungicides are used to kill or inhibit growth of fungi or insects that parasitize crops [32]. Pesticides originating from human activity can also enter water bodies through surface runoff, leaching, and/or erosion. Pesticides can cause endocrine disruptions and neurological disturbancies, influence immune system, reproduction and development [33].

Wastewaters containing dyes or pesticides are often treated by conventional methods as ozonation, membrane filtration, reverse osmosis, oxidation, ion exchange coagulation and adsorption [34–38]. Adsorption is considered as an attractive and favorable alternate for the removal of dyes and other organic molecules from wastewater streams. Many efforts have been made to find an appropriate adsorbent. Clay adsorbents enable to adsorb anionic and cationic pollutants such as dyes, pesticides and metal ions.

In this context, the present chapter focused in the application of Mostaganem bentonite in the treatment of natural water contaminated by dyes and pesticide. We will also study the phenomenon of adsorption of these pollutants through the use of mathematical models to determine the reaction mechanism, kinetics and thermodynamics of adsorption.

## 2. Materials and methods

Bentonite used in this investigation was purchased from M'zila deposit (Mostaganem, Algeria). This material is commercialized as industrial charge bentonite without additives by ENOF Company. The physicochemical properties of bentonite are resumed in chemical composition, point of zero charge, cation exchange capacity (CEC) and different analyses techniques such as XRD, FTIR, SEM and Specific surface area (BET).

The aim of the present study was the adsorption of two dyes, acid yellow E-4G and reactive yellow MX-4R. For this purpose two adsorbents were used: (i) bentonite intercalated by hydroxy-aluminum cations, (ii) bentonite pillared by cethyltrimethyl ammonium cation.

Bemacid Yellow E 4G (C.I. acid Yellow 49) and reactive Procion Yellow MX 4R (C.I. Reactive Yellow 14) were supplied by SOITEX textile society (Tlemcen, Algeria). The chemical formula and molecular weight of E-4G and MX-4R are $C_{16}H_{13}C_{12}N_5O_3S$; 426.24 g/mol and $C_{20}H_{19}ClN_4Na_2O_{11}S_3$; 669.02 g/mol, respectively. A stock solution (1.0 g/L) of dye was prepared by dissolving accurate weight amount in deionized water and the other concentrations were obtained by dilution of this stock dye solution.

The product chlorothalonil (Chl) fungicide was removed by acid activated bentonite from aqueous solution. Chl was obtained by Syngenta Protection of Plants S.A, Bale, Switzerland. It contains 400 g/L of chlorothalonil in the form of concentrated suspension with some impurities. The Chl is an inhibitor of spore germination, which acts on various enzymes and on the metabolism of fungi. The chemical formula of Chl is $C_8Cl_4N_2$, and its molecular weight is 265.93 g/mol. The solubility of chlorothalonil in water is 0.6 mg/L at 20 °C.

## 2.1 Preparation of pillared bentonite

The Al(III)-modified bentonite was obtained by mixing $AlCl_3$ solution (0.2 mol/L) with sodium hydroxide solution (0.2 mol/L) at 60 °C, up to the molar ratio $OH^-/Al^{3+} = 2$ [39]. The solution was aged at room temperature for three days before using. The resulting pillaring solution was added to the bentonite by stirring for 4 h at 70 °C at the ratio of 50 mmol oligomeric cations per gram of bentonite [40]. The slurry was stirred for 24 h at room temperature, filtered, and washed repeatedly with deionized water. The solid was dried at 80 °C and kept in a sealed bottle. The pillared bentonite obtained was designated as B-Al.

## 2.2 Preparation of CTAB-intercalated bentonite

The surfactant CTAB intercalated bentonite was synthesized as follows: the amount of CTAB (175 mg) corresponding to 1.0 times CEC of bentonite was dissolved in 1 L of distilled water at ambient temperature and stirred for 24 h. A total of 1 g bentonite was added to 100 mL surfactant solution. The dispersion was stirred for 4 h at 60 °C. The separated sample was washed several times and dried at 80 °C. The final product was noted as B-CTAB.

## 2.3 Acid activation of bentonite

Raw bentonite was treated by hydrochloric acid (HCl) (37% purity, Merck) at different concentrations (0.1, 1 and 6 N) at 70 °C. The amounts of 4 g of each treated sample were added to 400 mL of acid solution [41]. The contact time of the samples with the acid solution was fixed as 4 hours. At the end of treatment, the bentonite was washed several times with distilled water and dried over night at 80 °C [42].

## 2.4 Adsorption experiment

The adsorption experiments were carried out in a series of Erlenmeyer flasks containing 0.1 g of B-Al or B-CTAB and 20 mL of dyes aqueous solution at the desired concentration and initial pH (adjusted with hydrochloric acid 0.1 N or

0.1 N NaOH) in ambient temperature bath (23 °C). After shaking for 3 h of contact time, the flasks were removed and the concentration of MX-4R and E-4G after the adsorption was analyzed by spectrophotometer at wavelength of 425 and 400 nm, respectively. The adsorbed amount of dye (mg/g) was calculated as follow:

$$q_e = (C_0 - C_e) \cdot \frac{V}{m} \tag{1}$$

where, $q_e$ (mg/g) is the equilibrium adsorption capacity, $C_0$ the initial dye concentration, $C_e$ the equilibrium dye concentration (mg/L), $V$ the volume of solution (L) $m$ is the mass of the adsorbent (g).

The chlorothalonil was prepared in the range of initial concentrations 100–500 mg/L, in order to know the maximum amount of fungicide that bentonite can adsorb. For each experiment, 20 mL of pesticide solution was added to 0.1 g of the solid. The suspension was shaken at room temperature (23 °C) for 3 h. The chlorothalonil was detected by spectrophotometer at wavelength of 360 nm.

## 3. Results and discussions

### 3.1 Characterization of raw and modified bentonite

The chemical analysis of raw bentonite was performed by X-fluorescence XRF 9900. Thermo Instrument. The result of this analysis revealed that the silica (64.22%), alumina (11.62%) and lime (9.33%) are the main oxides of the bentonite with the existence of the others oxides in the small amounts such as $Fe_2O_3$ (4.88%), $TiO_2$ (1.06%), $Na_2O$ (3.38%) and $P_2O_5$ (0.03%). The elementary analysis of M'zila bentonite gives the formula $Na_{0.13}$, $Ca_{0.01}$, $K_{0.10}$, ($Al_{1.24}$ $Mg_{0.2}$ $Fe_{0.17}$ $Ti_{0.01}$) ($Si_{4.24}$) $O_{10}$ $(OH)_2$ with mass molar 368.68 g/mol [43].

The $pH_{PZC}$ value purified bentonite was found as 6.8. This value informs us about the electric charge on the solid surface. At pH value below than $pH_{PZC}$ the electrical charge on the surface is positive, it will be negative at pH higher than $pH_{PZC}$. Cation exchange capacity of natural bentonite was determined to be 112 meg/100 g by applying the conductimetric titration method [44]. The BET specific surface area measured via Quantachrome instrument was found 59.02 $m^2$/g.

The diffractogram of raw bentonite was shown in **Figure 2**. The bentonite sample contains some mineral phases such as the Montmonrillonite (M), Kaolinite (K), Calcite (C), Quartz (Q) and Dolomite (D). The characteristic peak $d_{001}$ of montmorillonite appears at 2θ = 5.5°, the kaolinite is observed at 2θ = 10° and the peak of calcite appears at 2θ = 31°.

The FT-IR spectrum of raw bentonite was performed with Agilent Cary 630 Spectrometer in range of 4000–400 $cm^{-1}$. The FT-IR analysis illustrated in **Figure 3** shows an intensive band at 1000 $cm^{-1}$ which is attributed to the Si-O in plan stretching vibration and other bands at 520 and 470 $cm^{-1}$ assigned to Al- O- Si (octahedral Al) and Si- O-Si bending vibrations, respectively [45]. The small band at 1620 $cm^{-1}$ is attributed to the deformation vibrations of the O–H bond of the constitution water. The band at 3620 $cm^{-1}$ is assigned to hydroxyl groups $Al^{3+}$ is partially replaced by $Fe^{3+}$ and $Mg^{2+}$.

The images of scanning electron microscope were made by microscope JSM-6360 to observe the morphology of bentonite particles. According to the **Figure 4** the particles were formed by heterogeneous aggregates of different shape and size. It appears that these grains constitute a stack of sheets probably representing the

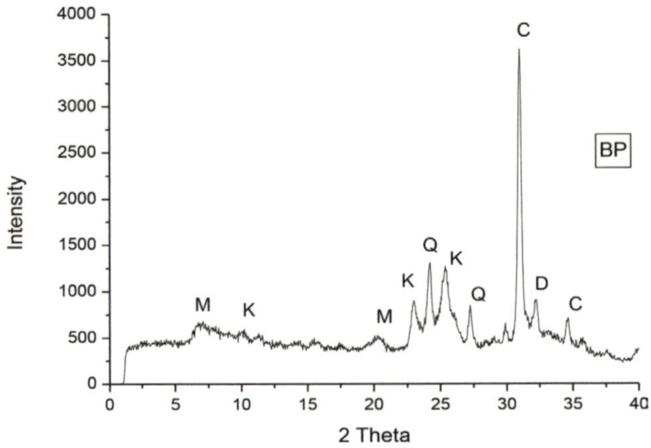


**Figure 2.**
*XRD pattern of raw bentonite.*

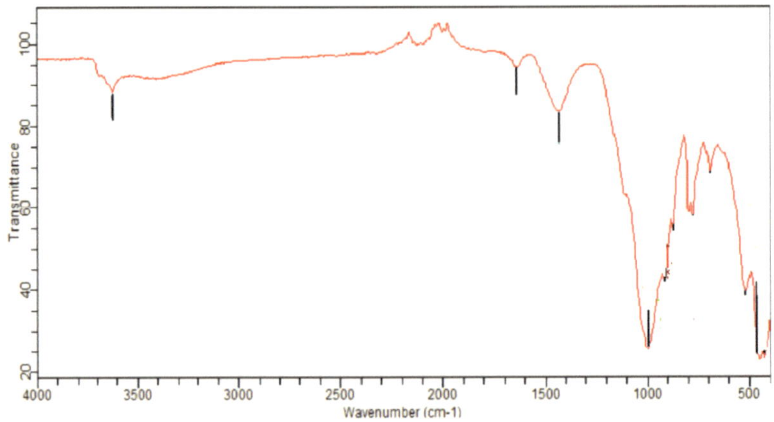


**Figure 3.**
*FT-IR spectra of raw bentonite.*

clay layers. On the surface of the sample, a small luminous crystallite settles, may be of free silica (quartz).

According to the XRD analysis realized with INEL CPS 120 instrument employing cobalt radiation ($\lambda$ = 0.178 nm), the hydroxy-aluminum polycations exchange increases the $d_{001}$ value to 14.3 Å, but the peak was much less intense compared to that of natural bentonite (**Figure 5**). The reduction in diffractogram might be caused by collapsing of the Mt. layers due to partial incongruent phase transition of hydroxy-Al into Fe/Al oxides and their interactions during aging and drying, as suggested by Thomas et al. [46].

The addition of surfactant causes the increasing of basal spacing of the bentonite around 18 Å, indicating location of $CTA^+$ ions between layers of montmorillonite. In order to increase more again the basal spacing, it must be increase in $CTA^+$ concentration, because as known, the amount of added surfactant has a direct effect on the interlayer expansion of Mt.

The pillaring bentonite with hydroxy-aluminum and CATB generated an enhancement of specific surface area where the values found were 110 and 194.4 $m^2/g$ for B-Al and B-CTAB, respectively. These values were very higher than that of untreated bentonite (59.02 $m^2/g$). In the case of Al-modified bentonite the specific surface area was increased significantly but a slight increase was noted

**Figure 4.**
*SEM images of natural bentonite extension 3000 and 9000.*

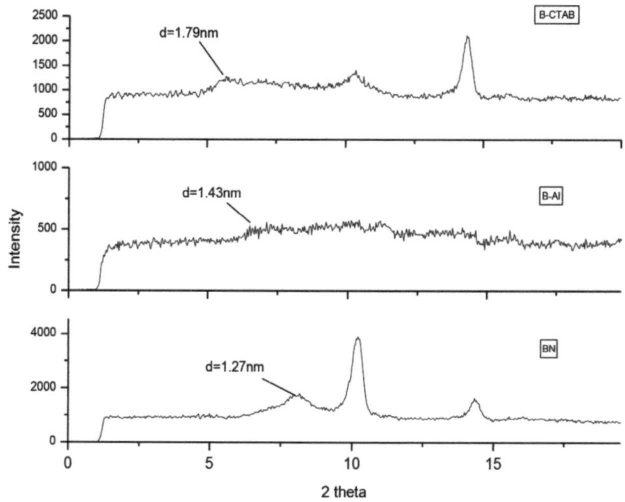


**Figure 5.**
*XRD patterns of natural and pillared bentonitew.*

through the basal spacing. This is due to the existence of electrostatic bonding between the negatively charges layers and pillaring oligocations in uncalcined Al-pillared clay.

The second application deals with the acid activation of Algerian bentonite and testing of his capacity to remove the chlorothalonil fungicide in aqueous solution. The hydrochloric acid solutions were used in the concentration range of 0.1-6 N.

The specific surface area of bentonite treated by 1 N (BA1N) and 6 N (BA 6 N) of hydrochloric acid were determined as 82.22 and 80.55 $m^2/g$, respectively. We see that the specific surface areas of both activated bentonites are almost identical, which are much higher than that of the raw bentonite (59.02 $m^2/g$). The specific surface area greatly increases at the acid concentration of 1 N, but slightly decreases at the concentrations higher than this value and then does not change much. Similar results were found by activating bentonite with sulfuric acid [47, 48].

The XRD patterns of raw and activated bentonites at various concentrations (0.1, 1 and 6 N) were shown in **Figure 6**. We note that there is no difference between the spectra of BN and BA 0.1 N. The concentration of 0.1 N of hydrochloric solution seems not be sufficient to make significant changes in the structure of

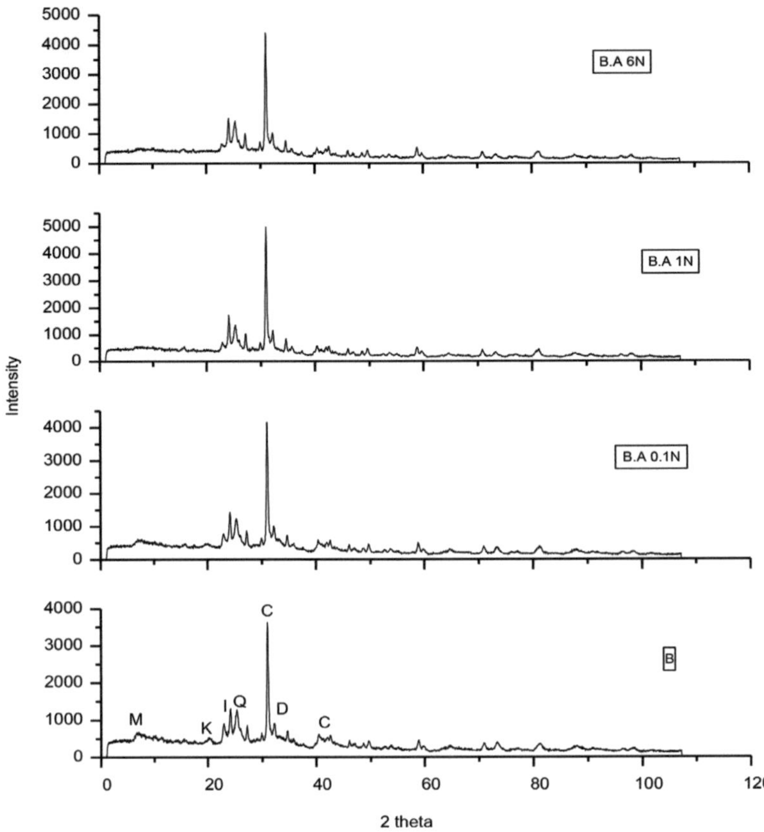


**Figure 6.**
*XRD patterns of raw and activated bentonites.*

bentonite. This means that it is a cation exchange causing the substitution of the exchangeable cations of interlayer space by protons $H^+$. In contrast to the sample treated with 0.1 N, the samples BA 1 N and BA 6 N undergo a significant structural modification according to the XRD spectra, where we notice that the peaks of montmorillonite and the kaolinite almost disappeared, while that of the illite is reduced in intensity. So from 1 N concentration of HCl, the clay minerals of bentonite are exposed to the direct effect of acid leading to the destruction of the basic clay sheets.

This process generally increases the surface area and the acidity of the clay minerals [49].

### 3.2 Adsorption isotherms of dyes and models fitting

The adsorption isotherms are realized at different initial concentrations of dyes, adsorbent dose was 5 g/L, and pH effect was tested and maximum adsorbed amount of dye was noted at pH = 2–3. The isotherms are formed by amount of dye adsorbed by the plot vs. equilibrium concentration. The adsorption isotherms of MX-4R and E-4G by the pillared bentonites are presented in **Figure** 7. The results show that the amount of dye adsorbed increases with the increase in equilibrium concentration of dye. All isotherms are of S-shape according to the classification of Giles et al. [50]. This type of isotherm originates from the cooperative isothermal adsorption, i.e. the adsorbed molecules promote higher adsorption of other molecules and tend to be adsorbed in groups. The maximum adsorbed amounts registered were 93.91 and

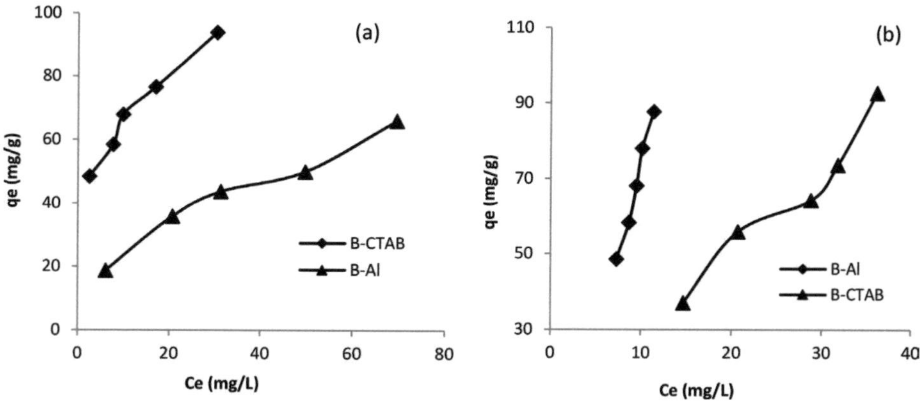


**Figure 7.**
*Adsorption isotherms of (a) MX-4R and (b) E-4G onto B-Al and B-CTAB. ($C_o$ = 200–500 mg/L, pH = 2–3, adsorbent dose 5 g/L, contact time 3 h)*

92.75 mg/g for MX-4R and E-4G respectively, attributed to B-CTAB sample. Those of B-Al adsorbent were 66.08 and 87.72 mg/g, respectively. According to these results, the B-CTAB sample has adsorption capacity most important than that of B-Al for the both dyes, in same operating conditions.

The adsorption isotherms were fitted by Langmuir and Freundlich equations expressed in linear forms in relations (2) and (3), respectively:

$$\frac{C_e}{q_e} = \frac{C_e}{Q0} + \frac{1}{K_L Q0} \qquad (2)$$

where $Q_0$ is the maximum adsorption capacity (mg/g), and $K_L$(L/mg) is a constant that relates to the heat of adsorption.

$$\log q_e = \log K_F + \frac{1}{n}\log Ce \qquad (3)$$

$K_F$ and $n$ are the Freundlich constants, indicating the capacity and intensity of adsorption, respectively.

The Langmuir and Freundlich constants and the linear regression correlations ($R^2$) for both isotherms model are listed in **Table 1**. The results reveal that the adsorption isotherms correlate with Freundlich model because the correlation coefficient ($R^2$) values obtained were above 0.98, while the model of Langmuir describes less well the experimental data where the linearization constants were insignificant except in the case of MX-4R. However Freundlich model is well used to describe the adsorption behavior of dyes on the both materials. This can be explained by the fact that the Langmuir equation is valid for monolayer adsorption onto a surface containing a finite number of identical sites, while Freundlich isotherm represents satisfactorily the sorption data on heterogeneous surfaces.

### 3.3 Adsorption kinetics of dyes

To evaluate the adsorption rate, the adsorption kinetic was examined by pseudo-first order and pseudo-second order models. The evolution of adsorption capacity of MX-4R and E-4G dyes increases over time and attained an equilibrium state around 60 min. The adsorption rates of the E-4G and MX-4R by intercalated bentonites are rapid early in the process and becoming slower over time. The

| Model | | Langmuir | | | Freundlich | | |
|---|---|---|---|---|---|---|---|
| Constants | | $Q_0$(mg/g) | $K_L$ (L/mg) | $R^2$ | 1/n | $K_F$ mg/g(L/mg)$^{1/n}$ | $R^2$ |
| B-Al | E-4G | ins | ins | ins | 1.396 | 2.96 | 0.984 |
| | MX-4R | 76.92 | 0.044 | 0.989 | 0.455 | 8.60 | 0.988 |
| B-CTAB | E-4G | ins | ins | ins | 0.920 | 3.177 | 0.955 |
| | MX-4R | 113.63 | 0.137 | 0.994 | 0.346 | 28.73 | 0.984 |

*ins: insignificant results.*

**Table 1.**
*Linearization constants of Langmuir and Freundlich equations.*

pseudo-first order kinetic using the linear Lagergren equation is generally expressed as follows [51]:

$$\ln(q_e - q_t) = \ln q_e - k_1 t \tag{4}$$

where $q_t$ is amount adsorbed of dye at time $t$ (mg/g) and $k_1$ is the rate constant of the pseudo-first order model (min$^{-1}$). The $k_1$ and $q_e$ were calculated from the slope and intercept of plots $\ln(q_e\text{-}q_t)$ versus $t$, respectively.

The pseudo-second order kinetic is expressed as follows [52]:

$$\frac{t}{q_t} = \frac{1}{k_2 q_e^2} + \frac{t}{q_e} \tag{5}$$

where $k_2$ is the rate constant of the pseudo-second order model for the adsorption process (g/mg.min). Plots of $t/q_t$ against $t$ have been drawn to obtain the rate parameters.

The calculated $q_e$ values agree with the experimental $q_e$ values, and the correlation coefficients for the pseudo-second-order kinetic plots were also found to be very high ($R^2$ = 0.98), indicating that pseudo-second order model fitted very well the kinetic adsorption. The pseudo second order model is based on the assumption that the rate limiting step may be chemisorption which involves valence forces by sharing or electron exchange between the adsorbent and the adsorbate [53]. According to the **Table 2**, the rate constants increase from 0.002 and 0.011 g.mg$^{-1}$ min$^{-1}$ (B-CTAB) to 0.922 and 0.912 g.mg$^{-1}$ min$^{-1}$ (B-Al). This means that the adsorption of dyes onto B-Al is a fast reaction, and CTAB-modified bentonite has a best adsorption capacity due to its high porosity and its large specific surface area.

### 3.4 Adsorption isotherms of chlorothalonil

The **Figure 8** shows the adsorption of Chl by the activated bentonite. This figure indicates that the adsorbed amount of Chl onto raw and activated bentonite increases in parallel with the equilibrium concentration. The experimental isotherm obtained here may be classified as type S referring to the classification of Giles et al. This type of isotherm originates from the cooperative isothermal adsorption, i.e. the adsorbed molecules promote higher adsorption of other molecules and tend to be adsorbed in groups. We note also that the activated bentonite adsorbs much better than the natural bentonite. When the activated

| | B-Al | | B-CTAB | |
|---|---|---|---|---|
| **Model** | **E-4G** | **MX-4R** | **E-4G** | **MX-4R** |
| Pseudo-first order | | | | |
| $q_{e(cal)}$ (mg/g) | 0.835 | 19.36 | 35.70 | 3.010 |
| $K_1$ ($min^{-1}$) | 0.019 | 0.065 | 0.043 | 0.019 |
| $R^2$ | 0.809 | 0.902 | 0.970 | 0.350 |
| Pseudo-second order | | | | |
| $q_{e(cal)}$ (mg/g) | 12.048 | 45.455 | 42.016 | 60.24 |
| $K_2$ (g/mg.min) | 0.922 | 0.912 | 0.002 | 0.011 |
| $R^2$ | 0.999 | 0.997 | 0.996 | 0.999 |
| $q_{e(exp)}$ (mg/g) | 11.865 | 44.207 | 38.926 | 59.588 |

**Table 2.**
*Constants rates of the E-4G and MX-4R adsorption by B-Al and B-CTAB.*

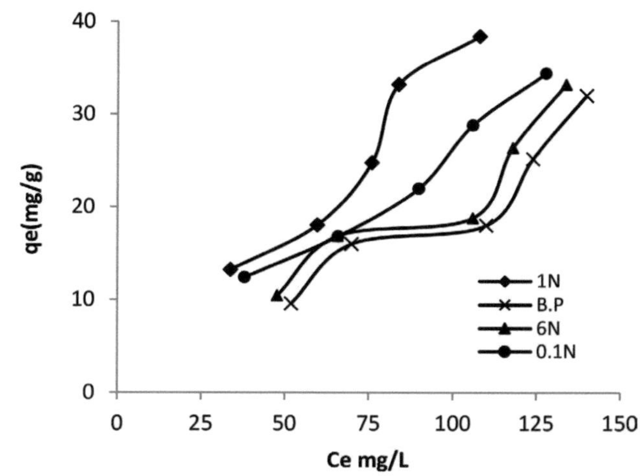


**Figure 8.**
*Adsorption isotherms of Chl onto activated bentonite.*

| **Sample** | **T (K)** | **$K_F$ (mg/g (L/mg)$^{1/n}$)** | **1/n** | **$R^2$** |
|---|---|---|---|---|
| BA 0.1 N | 296 | 0.913 | 0.737 | 0.935 |
| BA 1 N | 296 | 0.590 | 0.939 | 0.925 |
| | 303 | 2.064 | 0.725 | 0.950 |
| | 313 | 2.295 | 0.831 | 0.979 |
| | 323 | 2.087 | 0.736 | 0.970 |

**Table 3.**
*The constants of Freundlich model.*

samples were compared, it was found that the maximum amount of adsorption (38.42 mg/g) was observed for BA 1 N.

The isotherm model fit the experimental data very well is Freundlich model. The Freundlich equation is an empirical equation that can be used for heterogeneous systems with interaction between the molecules adsorbed. As seen from **Table 3**, the values of regression coefficients $R^2$ were close to the unit and the 1/n values were less than unity which indicates that the adsorption intensity was favorable and was a physical process. On the other hand, the experimental data fitted by Langmuir model were insignificant in terms of adsorption; this is why they are not mentioned.

### 3.5 Adsorption heats of chlorothalonil

In the case of adsorption of molecules on a solid surface, the Gibbs energy is composed of two functions, the enthalpy function ($H$), which is measure of the energy of interaction between the molecules and the adsorbent surface, and the entropy function ($S$), which reflects the change and the arrangement of molecules in the liquid phase and on the surface. The free energy $\Delta G$ was calculated according the following relation:

$$\Delta G^0 = \Delta H^0 - T\Delta S^0 \quad (6)$$

The distribution coefficient $K_d$ (L/g) is calculated from the following Equations [54, 55]:

$$\ln K_d = \frac{\Delta S^0}{R} - \frac{\Delta H^0}{RT} \quad (7)$$

$$K_d = \frac{(C_0 - C_e)}{C_e} \cdot \frac{V}{m} \quad (8)$$

where $\Delta H°$, $\Delta S°$, and $T$ are the adsorption enthalpy (kJ/mol), entropy (J/mol.K) and temperature in Kelvin, respectively, and $R$ is the gas constant (8.31 J/mol.K). The slope and intercept of the plot of ln $K_d$ versus 1/T correspond to $\Delta H°/R$ and $\Delta S°/R$, respectively.

It can be seen from **Table 4** that the calculated $\Delta G°$ values have negative signs, indicating that the adsorption process is spontaneous in the experimental conditions and the spontaneity increases with increasing of temperature. All enthalpy values are positive, showing that the adsorption process is endothermic. The magnitude of enthalpy values suggests that the adsorption of Chl onto activated bentonite was physic in nature. The entropy values were positive that means the molecules disorder was located in interface solution/solid.

| | $C_0$(mg/L) | ΔH° (kJ/mol) | ΔS° (J/molK) | ΔG° (kJ/mol) | | | |
|---|---|---|---|---|---|---|---|
| | | | | 303 K | 313 K | 323 K | $R^2$ |
| BA 1 N | 30 | 68.894 | 234.292 | - 2.096 | - 4.439 | - 6.782 | 0.930 |
| | 40 | 34.020 | 120.528 | - 2.499 | - 3.705 | - 4.910 | 0.895 |
| | 50 | 29.726 | 107.381 | - 2.810 | - 3.883 | - 4.957 | 0.927 |
| | 60 | 39.861 | 137.879 | - 1.915 | - 3.294 | - 4.673 | 0.926 |
| | 80 | 40.422 | 141.319 | - 2.397 | - 3.810 | - 5.223 | 0.910 |

**Table 4.**
*Heats adsorption of Chl onto activated bentonite.*

## 4. Conclusions

In this research we studied the characteristics and the physicochemical properties of an Algerian bentonite which its adsorption capacity was tested to eliminate organic pollutants from aqueous solution.

Before the adsorption experiment, bentonite underwent two different treatments in order to improve its exchange capacity and porosity. The first treatment carried out is that of the pillaring clay with mineral (polycations of $Al_{13}$) and organic (CTAMB) intercalants. The second is the treatment by acid attack (HCl) at different concentrations (0.1, 1 and 6 N).

The intercalation of the bentonite by Al13 and CTAB increased the basal sheet space up to 14.3 and 18 Å, respectively. The chemical activation with HCl at 6 N concentration enhanced the specific surface area of the bentonite from 59.02 to a value of 82 $m^2/g$. The obtained materials from the both treatments were applied for the adsorption of MX-4R, E-4G dyes and the fungicide chlorothalonil.

The adsorption isotherms of these pollutants have shown that the adsorption capacities were very satisfactory and the adsorption phenomenon was physical nature. The adsorption isotherms of all adsorbates were well described by Freundlich model. Kinetic data of dyes adsorption tend to fit well in pseudo-second order rate expression. Moreover the adsorption of chlorothalonil by activated bentonite was spontaneous and this spontaneity increases with increasing temperature.

## Conflict of interest

The authors declare no conflict of interest in publishing this chapter.

## Author details

Reda Marouf*, Nacer Dali, Nadia Boudouara, Fatima Ouadjenia and Faiza Zahaf
Faculty of Exacts Sciences, Laboratory of Materials, Applications and Environment, University Mustapha Stambouli of Mascara, Algeria

*Address all correspondence to: r.marouf@univ-mascara.dz

## References

[1] Achour S, Youcef L, élimination du cadmium par adsorption sur bentonites sodique et calcique (Cadmium removal by adsorption on sodic and calcic bentonites) . Larhyss Journal. 2003;2:69-81.

[2] Youcef L, Achour S, Etude de l'élimination des fluorures des eaux de boisson par adsorption sur bentonite (Removal study of fluorides from drinking wtaers by adsorption on bentonite). Larhyss Journal. 2004; 3:129-142.

[3] Suquet H, de la Calle C, Pezerat H, Swelling and structural organization of saponite. Clay Clay Miner. 1975;23: 1-9. DOI: 10.1346/CCMN.1975.0230101

[4] Arfaoui, S, Frini-Srasra N, Srasra E, Application of clays to treatment of tannery sewages. Desalination. 2005; 185 :419-426. DOI: 10.1016/j.desal.2005.04.047

[5] Hamdi, N, Srasra E, Remove of fluoride from acidic wastewater by clay mineral effect of solid–liquid ratios. Desalination. 2007; 206:238-244. DOI: 10.1016/j.desal.2006.04.054

[6] Al-Asheh S, Banat F, Abu-Aitah L, Adsorption of phenol using different types of activated bentonites. Sep. Purif. Technol. 2003; 33: 1-10. DOI : 10.1016/S1383-5866(02)00180-6

[7] Fu-Chuang H, Jiunn-Fwu L, Chung-Kung L, Huang-Ping C, Effects of cation exchange on the pore and surface structure and adsorption characteristics of montmorillonite. Colloid Surf. A: Physicochem. Eng. Aspects. 2004; 239: 41-47. DOI: 10.1016/j.colsurfa.2003.10.030

[8] Min-Yu T, Su-Hsia L, Removal of basic dye from water onto pristine and HCl activated montmorillonite in fixed beds. Desalination. 2006;194: 156-165. DOI : 10.1016/j.desal.2005.11.008

[9] Okada, K, Arimitsu N, Kameshima Y, Akira N, Kenneth MacKenzie JD, Solid acidity of 2:1 type clay minerals activated by selective leaching. Appl. Clay Sci. 2006;31: 185-193. DOI: 10.1016/j.clay.2005.10.014

[10] Pinnavaia TJ, Intercalated clay catalysts. Science. 1983;220: 365-371. DOI: 10.1126/science.220.4595.365

[11] Carrado KA, Synthetic organo- and polymer-clays: preparation, characterization, and materials applications. Appl. Clay Sci. 2000;17: 1-23. DOI: 10.1016/S0169-1317(00)00005-3

[12] A. Vaccari, Preparation and catalytic properties of cationic and anionic clays. Catal. Today. 1998;41: 53-71. DOI: 10.1016/S0920-5861(98)00038-8

[13] Hernando MJ, Pesquera C, Blanco C, Gonzalez F, Comparative study of the texture of montmorillonites pillared with aluminum and aluminum/cerium. Langmui. 2001;17 : 5156-5159. DOI: 10.1021/la010157k

[14] Ding Z, Kloprogge JT, Frost RL, Porous clays and pillared clays-based catalysts. Part 2: a review of the catalytic and molecular sieve applications. J. Porous Mater. 2001;8: 273-293. DOI: 10.1023/A:1013113030912

[15] Gil A, Korili SA, Trujillano R, Vicente MA, A review on characterization of pillared clays by specific techniques. Appl. Clay Sci. 2011;53: 97-105. DOI: 10.1016/j.clay.2010.09.018

[16] Yan LG, Shan X-Q, Adsorption of cadmium onto Al13-pillared acid-activated montmorillonite. J. Hazard. Mater. 2008;156: 499-508. DOI: 10.1016/j.jhazmat.2007.12.045

[17] Tomul F, Adsorption and catalytic properties of Fe/Cr-pillared bentonites.

Chem. Eng. J. 2012;185-186: 380-390. DOI/ 10.1016/j.cej.2012.01.094

[18] Guerra DL, Airoldi C., Lemos VP,. Angélica RS, Adsorptive, thermodynamic and kinetic performances of Al/Ti and Al/Zr-pillared clays from the Brazilian Amazon region for zinc cation removal. J. Hazard. Mater. 2008;155: 230-242. DOI: 10.1016/j.jhazmat.2007.11.054

[19] Zhou J, Wu P, Dang Z, Zhu N, Li P, Wu J, Wang X, Polymeric Fe/Zr pillared montmorillonite for the removal of Cr(VI) from aqueous solutions. Chem. Eng. J. 2010;162: 1035-1044. DOI: 10.1016/j.cej.2010.07.016

[20] de Paiva LB, Morales AR, Diaz FRV, Organoclays: properties, preparation and applications. Appl. Clay Sci. 2008 ;42(1-2): 8-24. DOI: 10.1016/j.clay.2008.02.006

[21] Zhou Q, He HP, Zhu JX, Shen W, Frost RL, Yuan P, Mechanism of p-nitrophenol adsorption from aqueous solution by HDTMA+ pillared montmorillonite-implications for water purification. J. Hazard. Mater. 2008.;154(1-3): 1025-1032. DOI: 10.1016/j.jhazmat.2007.11.009

[22] Ozcan A, Omeroglu C, Erdogan Y, Ozcan AS, Modification of bentonite with a cationic surfactant: an adsorption study of textile dye Reactive Blue 19. J. Hazard. Mater. 2007;140(1-2): 173-179. DOI: 10.1016/j.jhazmat.2006.06.138

[23] Jarraya I, Fourmentin S, Benzina M, Bouaziz S, VOC adsorption on raw and modified clay materials. Chem. Geol. 2010;275 (1-2): 1-8. DOI: 10.1016/j.chemgeo.2010.04.004

[24] Park SJ, Kim YB, Yeo SD, Vapor adsorption of volatile organic compounds using organically modified clay. Sep. Sci. Technol. 2008; 43(5): 1174-1190. DOI: 10.1080/01496390801910138

[25] Jiang JQ, Cooper C, Ouki S, Comparison of modified montmorillonite adsorbents-part I: preparation, characterization and phenol adsorption. Chemosphere. 2002;.47(7): 711-716. DOI: 10.1016/S0045-6535(02)00011-5

[26] Zhu LZ, Zhu RL, Simultaneous sorption of organic compounds and phosphate to inorganic–organic bentonites from water. Sep. Purif. Technol. 2007.;54 (1): 71-76. DOI: 10.1016/j.seppur.2006.08.009

[27] Falaras P, Kovanis I, Lezou F, Seiragakis G, Cottonseed oil bleaching by acid activated montmorillonite. Clays Clay Miner. 1999;34: 221-232. DOI: 10.1180/000985599546181

[28] Hussin F, Aroua MK., Daud WMAW,. Textural characteristics, surface chemistry and activation of bleaching earth: a review. Chem. Eng. J. 2011;170: 90-106. DOI: 10.1016/j.cej.2011.03.065

[29] Oyewo OA, Elemike EE, Onwudiwe DC, Onyango MS, Metal oxide-cellulose nanocomposites for the removal of toxic metals and dyes from wastewater. Int. J. Biol. Macromol. 2020;164:2477-2496. DOI: 10.1016/j.ijbiomac.2020.08.074

[30] Kaemkit C, Monvisade P, Siriphannon P, Nukeaw J, Water soluble chitosan intercalated montmorillonite nanocomposites for removal of Basic Blue 66 and Basic Yellow 1 from aqueous solution. J. Appl. Polym. Sci. 2013;128(1): 879-887. DOI: 10.1002/app.38255

[31] Bolognesi C, Merlo FD, Pesticides: human health effects. In: Nriagu JO, editor. Encyclopedia of environmental health. Burlington: Elsevier; 2011. 438-453 pp. DOI: 10.1016/b978-0-444-52272-6.00592-4

[32] Gupta PK, Herbicides and fungicides. In: Gupta RC, editor.

Reproductive and developmental toxicology. Amsterdam: Academic Press/Elsevier; 2011. 503-521 pp. DOI: 10.5688/ajpe77120

[33] Khan ZM, Law FCP, Adverse effects of pesticides and related chemicals on enzyme and hormone systems of fish, amphibians and reptiles: A review. Proc Pakistan Acad Sci. 2005;42:315-323.

[34] Lima Beluci NC, Affonso Pisano GM, Sayury Miyashiro C, Cândido Homem N, Guttierres Gomesd R, Fagundes-Klenc MR, Bergamasco R, Marquetotti Salcedo VA. Hybrid treatment of coagulation/ flocculation process followed by ultrafiltration in $TiO_2$-modified membranes to improve the removal of reactive black 5 dye. Sci. Total Environ. 2019;664:222-229. DOI: 10.1016/j.scitotenv.2019.01.199

[35] Pradhan SS, Konwar K, Ghosh TN, Mondal B, Sarkar SK, Deb P, Multifunctional Iron oxide embedded reduced graphene oxide as a versatile adsorbent candidate for effectual arsenic and dye removal. Colloid Interface Commun. 2020;39: 100319. DOI; 10.1016/j.colcom.2020.100319

[36] Alvi MA, Shaheer Akhtar M, Effective photocatalytic dye degradation using low temperature grown zinc oxide nanostructures. Mater. Letters. 2020;281: 128609. DOI: 10.1016/j.matlet.2020.128609

[37] Mehta R, Brahmbhatt H, Mukherjee M, Bhattacharya A, Tuning separation behavior of tailor-made thin film poly(piperazine-amide) composite membranes for pesticides and salts from water. Desalination. 2017 ;404: 280-290. DOI: 10.1016/j.desal.2016.11.021

[38] Mohammadi P, Sheibani H, Evaluation, of the bimetallic photocatalytic performance of Resin–Au–Pd nanocomposite for degradation of parathion pesticide under visible light. Polyhedron. 2019;170: 132-137. DOI: 10.1016/j.poly.2019.05.030

[39] Roca Jalil ME, Vieira R, Azevedo D, Baschini M, Sapag K, Improvement in the adsorption of thiabendazole by using aluminum pillared clays. Appl. Clay Sci. 2013;71: 55-63. DOI: 10.1016/j.clay.2012.11.005

[40] Romero-Pérez A, Infantes-Molina A, Jiménez-Lopez A, Jalil ER, Sapag K, Rodriguez-Castellon E, Al-pillared montmorillonite as a support for catalysts based on ruthenium sulfide in HDS reactions. Catal. Today. 2012;187: 88-96. DOI: 10.1016/j.cattod.2011.12.020

[41] Steudel A, Batenburg LF, Fischer HR, Weidler PG, Emmerich K, Alteration of non-swelling clay minerals and magadiite by acid activation. Appl. Clay Sci. 2009; 44: 95-104. DOI: 10.1016/j.clay.2009.02.001

[42] Eloussaief M, Benzina M, Efficiency of natural and acid-activated clays in the removal of Pb(II) from aqueous solutions. J. Hazard. Mater. 2010;178: 753-757. DOI: 10.1016/j.jhazmat.2010.02.004

[43] Medout-Marere V, Belarbi H, Thomas P, Morato F, Giuntini JC, Douillard JM, Thermodynamic analysis of the immersion of a swelling clay. J. Colloid Interface Sci. 1998; 202: 139-148. DOI: 10.1006/jcis.1998.5400

[44] Navia R. Environmental use of volcanic soil as natural adsorption material [PhD Thesis]. University of Leoben (Austria); 2004.

[45] Eren E, Afsin B, Onal Y, Removal of lead ions by acid activated and manganeseoxide-coated bentonite. J. Hazard. Mater. 2009;161:677-685. DOI: 10.1016/j.jhazmat.2008.04.020

[46] Thomas SM, Bertrand JA, Occelli ML, Huggins F, Gould SA, MicroporousMontmorillonitesExpanded

with Alumina Clusters and M[(μ-OH) Cu (μ-$OCH_2CH_2NEt_2$)]$_6$ ($ClO_4$)$_3$, (M = Al, Ga, and Fe), or Cr[(μ-$OCH_3$) (μ-$OCH_2CH_2NEt_2$) CuCl]$_3$ Complexes. Inorg. Chem. 1999;38(9): 2098-2105. DOI: 10.1021/ic981127g

[47] Tomić ZP, Logar VP, Babic BM, Rogan JR, Makreski P, Comparison of structural, textural and thermal characteristics of pure and acid treated bentonites from Aleksinac and Petrovac (Serbia). Spectrochimica Acta Part A 2011;82: 389-395. DOI: 10.1016/j.saa.2011.07.068

[48] Caglar B, Afsin B, Koksal E, Tabak A, Eren E, Characterization of Unye bentonite after treatment with sulfuric acid. Quim. Nova. 2013; 36: 955-959. DOI: 10.1590/S0100-40422013000700006

[49] Bhattacharyya KG, Gupta SS,. Adsorption of a few heavy metals on natural and modified kaolinite and montmorillonite: A review. Adv. Colloid Interface Sci. 2008;140: 114-131. DOI: 10.1016/j.cis.2007.12.008

[50] Giles CH, MacEwan TH, Nakhwa SN, Smith D, Studies in adsorption. Part XI: A system of classification of solution adsorption isotherms, and its use in diagnosis of adsorption mechanisms and in measurements of specific surface areas of solids. J. Chem. Soc. 1960;10: 3963-3973.

[51] Ho YS, Citation review of Langergren kinetic rate equation on adsorption reactions. Scientometrics. 2004;59 : 171-177. DOI: 10.1023/B:SCIE.0000013305.99473.cf

[52] Ho YS, McKay G, The kinetics of sorption of divalent metal ions onto sphagnum moss peat. Water Res. 2000;34: 735-742. DOI: 10.1016/S0043-1354(99)00232-8

[53] Komadel P, Chemically modified smectites. Clay Miner. 2003;38: 127-138. DOI: 10.1180/0009855033810083

[54] Fan QH, Shao DD, Hu J, Wu WS, Wang XK, Comparison of $Ni^{2+}$ sorption to bare and ACT-graftattapulgites: effect of pH, temperature and foreign ions. Surf. Sci. 2008;602: 778-785. DOI: 10.1016/j.susc.2007.12.007

[55] Wang XK, Chen CL, Hu WP, Ding AP, Xu D, Zhou X, Sorption of $^{243}$Am(III) to multiwall carbon nanotubes. Environ. Sci. Technol, 2005;39: 2856-2860. DOI: 10.1021/es048287d

Section 3

# Montmorillonite Composite

Chapter 3

# Reinforcement of Montmorillonite Clay in Epoxy/Unsaturated Polyester Blended Composite: Effect on Composite Properties

*Chakradhar V.P. Komanduri*


## Abstract

Montmorillonite (MMT) clay was disseminated into Unsaturated Polyester (UP) and Epoxy blend systems in diverse weight ratios namely, 0, 1, 2, 3, and 5% to prepare Epoxy/UP/MMT clay composite. The specimen was characterized by thermal and chemical analysis. Homogeneous mixture of blended composites is obtained through mechanical stirring and ultrasonication processes. The testing of thermal and chemical properties was performed. Evidence acquired from the above tests indicate that Epoxy reinforced with UP and further strengthened with MMT clay enhanced the thermal and chemical properties of the composite to a considerable extent. The purpose of this study was to recognize an appropriate composite offering a stronger material with enhanced performance; that is suitable for diverse industrial uses.


**Keywords:** montmorillonite clay, epoxy, unsaturated polyester, thermal properties, chemical properties

## 1. Introduction

Polymer blending attained an appreciable market, since they save by weight approximately 36–40% of polymer consumption. Epoxy is a flexible, popular resin used for manufacturing state-of-art composites since it has superior binding, thermal, mechanical and aging characteristics [1, 2]. But for enhancement of impact attribute in state-of-art engineering uses, reinforcing epoxy is required. It can be improved through mixing with adaptable polymers and elastomers. Nevertheless, alteration of epoxy with elastomers enhances its toughness property besides drop in few properties of epoxy at elevated temperatures [3–5]. Hence, an apt polymer is needed to improve the toughness of epoxy, by preserving stiffness, glass transition temperature, and heat stability. It is accomplished by an inter cross-linked polymer network of thermoset-thermoset blends [6–10].

A good range of commercial relevance to polyester resins is identified in areas like paints and surface coatings. Several merits of polyester are observed with inclusion of flexibility in properties, reasonable cost and ease of use. But polyester possesses some demerits such as inferior resistance to alkalis and hardness. For overcoming the above-mentioned demerits, blending of polyester with suitable resins can be performed as it exhibits better compatibility with diverse resins. The technique of blending can be productively applied to eliminate the substandard properties of both components. Blended polymers provide superior composite from lesser superior components.

The merits of nano-particles over conventional macro or micro particles are improved surface area and aspect ratio which could improve binding of nano-particles and polymers. Presently the most popular clay used for polymer-clay composites is MMT. Various studies were conducted on epoxy-clay composites (ECN) under different curing conditions. The exfoliated clay provides superior properties and offers advantages over other nanofillers in terms of cost and biodegradability [8, 9]. This paper furnishes facts on the property's analysis of the MMT clay reinforced polymer blended composites. The intention of this study was to recognize a composite that offers better strength, providing better performance at minimum cost; applicable for diverse applications.

Composites are influenced by chemical and thermal properties. Thermal and chemical properties of materials play an equally important role as mechanical properties. Polymers are very vulnerable to changes in temperature. Plastics tend to become rigid and brittle at minimal temperatures. The portability of the polymer chain is greatly reduced at low temperatures, which is the reason for the above. The temperature and dimensions of solid increases as it absorbs heat. Further heating melts the solid. Thus, knowledge of heat properties of materials becomes crucial for assessing the performance of polymers and their reaction to thermal changes.

## 2. Experimental procedure

### 2.1 Materials and methods

i. **Epoxy** (Araldite-LY 556 and Amine hardener HY-951; ratio-100:10)

ii. **Unsaturated polyester** (Ecmalon 9911; accelerator: 2% cobalt naphthenate, catalyst: 2% methyl ethyl ketone peroxide (MEKP) and promoter: 10% dimethylaniline (DMA) solution), are the resin materials. Exfoliated MMT clay treated with 25% trimethyl stearyl ammonium is the reinforcement material (**product name:** hydrophilic bentonite powder; **Density:** 600–1100 kg/$m^3$; **Size:** 25 μm).

Thermogravimetric analysis (TGA) and Differential scanning calorimetry (DSC-2010 TA Instrument) are employed to assess the heat attributes of the epoxy/polyester/MMT clay blended composites [11]. The thermal degradation behavior of the composite blend is investigated using TGA. TGA is performed on a 10 mg powdered sample to identify changes in weight to corresponding changes in temperature. The sample is placed in a thermocouple fitted oven for precise temperature measurement. An inert gas atmosphere is used to suppress unwanted reactions. A computer monitors and controls the entire process. Steadily enhancing sample temperature until 1200°C, evaluation is performed. A graph of percentage weight Vs temperature is plotted [12].

Evaluation of Glass transition temperature ($T_g$) of the material and assessment of thermal deterioration of polymers is carried out using DSC [13]. Studies are performed under an inert atmosphere at 10°C/min scan rate and a temperature ranging from 30°C up to 600°C. A 10 mg powdered specimen is used for each run. Thermograms are plotted recording weight change with respect to temperature change. Weight change results from bond-forming or breaking at elevated temperatures. Throughout the experiment identical temperatures are maintained for the sample and reference. Heat capacity over a span of temperatures was examined. When the sample undergoes phase transition, heat is required to flow to it and the reference to sustain them at similar temperatures. Suppose as a solid sample changes phase to liquid it needs more heat flow to enhance its temperature at the same pace as the reference. DSC can estimate quantity of heat taken in and discharged in sample and reference by perceiving variation in heat flow. DSC can also perceive precise change of state, like glass transitions. It is extensively applied in industries as a quality checking instrument to estimate sample clarity and to observe curing of polymers.

To analyze the chemical resistance of the composites, ASTM 543–87 test methodology is employed. Standard reagents are applied to confirm outcomes. In the existing work, tests on epoxy/UP/clay composites are conducted to identify the resistivity to chemicals. The reaction of acids, alkalis and solvents on the composite are studied. In each case, pre-weighed specimens (5 x 5 x 3 mm) of 10 numbers were dipped into the respective chemical reagents for twenty-four-hour duration.

## 2.2 Thermal properties

The thermal properties analysis like TGA and DSC are performed on clay-filled epoxy/UP blended composite as per ASTM E1131 and ASTM D3418 respectively. The properties like degradation temperature and glass transition temperature are studied. In TGA, the heat stability of composite is studied as % weight loss $v_{s.}$ temperature.

## 2.3 Resistivity to chemicals

ASTM 543–87 is employed to evaluate the chemical resistivity of the composite. In the present work, chemical resistivity tests on epoxy/UP/clay composites are conducted. In each case, ten specimens are tested. Pre-weighed specimens are dipped in chemical reagents for 24 hours. They are then removed from the chemicals, cleaned in distilled water, and completely dried at room temperature using filter paper. The specimens are again weighed and the % gain/decrease in weight is ascertained as shown below.

$$\%\text{gain / decrease in specimen weight} = \frac{\text{Original weight} - \text{Final weight}}{\text{Original weight}} \times 100 \qquad (1)$$

The chemical reagents used in the study are mentioned below:
*Acids:* Acetic acid, Nitric acid and Hydrochloric acid.
*Alkalis:* Sodium hydroxide, Sodium carbonate and Ammonium hydroxide.
*Solvents:* Benzene, Toluene, Carbon tetrachloride and Water.

## 2.4 Results and discussions

### *2.4.1 Thermogravimetric analysis*

Heat stability of composites is analyzed using TGA. **Figure 1** indicates the weight loss of five different samples. In pure blend, loss in weight is constant up to 200°C and degradation starts at 400°C. As clay content is increased to 5 wt. % degradation temperature of the composite shifts upwards to higher temperatures. Until 350°C Weight loss is constant for 5 wt. %. Presence of moisture is the cause

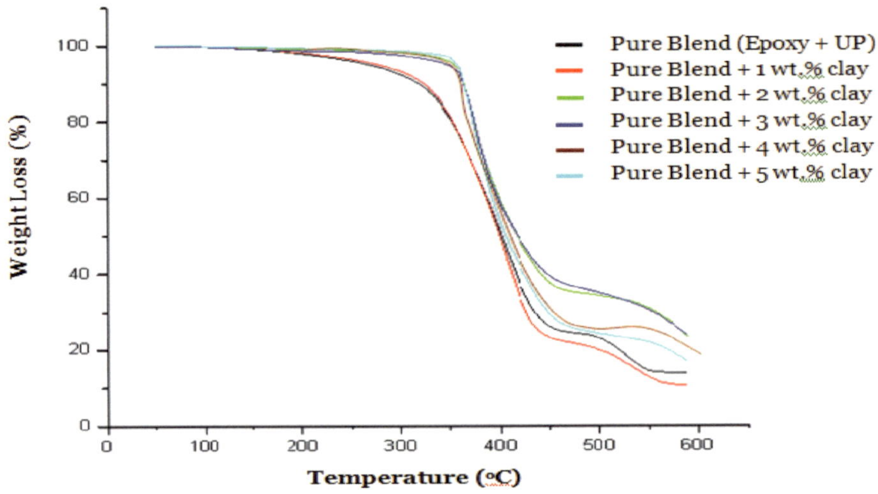


**Figure 1.**
*Epoxy/polyester as a function of MMT clay-TGA.*

for drop in weight for 1 wt. % sample. Evidently, the degradation temperature of the composite shifts upwards to elevated temperatures, demonstrating enhanced heat stability for 5 wt. % clay [12, 13]. Presence of inorganic materials like clay is the cause for improved heat stability.

In comparison to pure blend 15% weight loss and a 10° C rise in degradation temperature are observed for 5 wt. % sample. The results indicate that better dispersion of polymer and clay are the reasons for higher heat stability [13].

### *2.4.2 Differential scanning calorimetry*

DSC investigates the thermal transitions of the pure polymer and the composites. A graph indicating glass transition temperatures for diverse clay weight percentages viz. for 0, 1, 2, 3, 4 and 5 wt.% at 430, 432, 433, 433, 431 and 429°C respectively of the composite is presented in **Figure 2**. A fall of 2°C glass transition temperature is noticed for 4 wt%., whereas a fall of 4°C is observed for 5 wt.% in comparison to 3 wt.%. Nil change is perceived for 2 and 3 wt. % samples. Similar changes in $T_g$ are due to: (i) better surface interaction strengthening the interface (ii) enhanced interfacial free volume due to the lower bulk crystallinity of polymer chains. It is also found that the $T_g$ of the Epoxy/UP composite decreases at higher montmorillonite loading.

### *2.4.3 Chemical resistance measurement*

From **Table 1** it is evident that the Epoxy/UP/MMT clay composite exhibits better resistivity to all chemicals considered for the study, except the solvents. In each case, the pre-weighed specimens are immersed in the chemicals, cleaned in distilled water and then dried. Nanocomposite blend specimens show a weight reduction on treatment with solvents. This is understandable as UP, that is present in the blend, dissolves in the solvents under study. These composites prove to be having good resistance to attack from chemicals except solvents. The highly expandable montmorillonite clay has caused maximum swelling in nitric acid in contrast to maximum weight loss in benzene due to its high cation exchange capacity. The increase in weight of the composite is due to the penetration of the liquid chemical resulting due to swelling of composite [14].

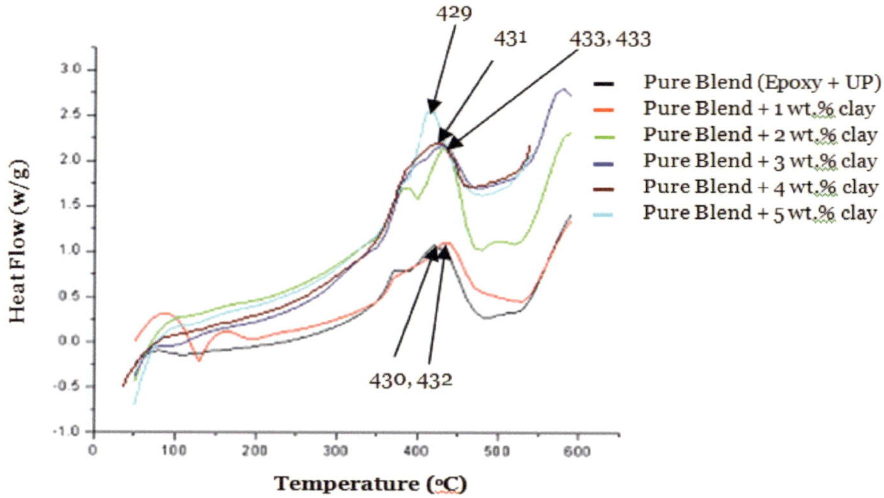


**Figure 2.**
*Epoxy/polyester blend as a function of MMT clay-DSC analysis.*

| Chemical used | Clay (wt. %) | | | | | |
|---|---|---|---|---|---|---|
| | 0% | 1% | 2% | 3% | 4% | 5% |
| HCl (10%) | +1.327 | +0.925 | +0.845 | +0.455 | +0.535 | +0.576 |
| $CH_3COOH$ (5%) | +1.256 | +0.287 | +1.348 | +0.934 | +0.457 | +0.223 |
| $HNO_3$ (40%) | +1.978 | +1.546 | +1.656 | +1.645 | +1.698 | +1.728 |
| NaOH (10%) | +1.323 | +1.625 | +0.302 | +0.785 | +0.645 | +0.575 |
| $Na_2CO_3$ (20%) | +0.326 | +0.320 | +0.250 | +0.167 | +0.152 | +0.144 |
| $NH_4OH$ (10%) | +0.825 | +0.676 | +0.532 | +0.440 | +0.487 | +0.532 |
| Benzene | −2.450 | −5.247 | −5.420 | −7.575 | −7.897 | −8.352 |
| Toluene | −1.054 | −2.926 | −8.265 | −14.956 | −10.567 | −8.345 |
| $CCl_4$ | −0.364 | −0.956 | −3.958 | −17.256 | −12.345 | −8.745 |
| Water | +0.909 | +1.325 | +9.352 | +11.526 | +10.457 | +9.348 |

**Table 1.**
*Experimental values, showing percentage change in weight of epoxy/UP/MMT clay composite*

## 2.5 Conclusions

In the experimental analysis of thermal and chemical resistivity of MMT clay strengthened composite, the following conclusions were made.

From TGA it was observed that clay content does not affect the heat stability at 5 wt.% in comparison to pure and other combinations blends. A weight loss of 15% and rise of 10°C in degradation temperature were noticed in the TGA analysis, while 4°C fall in $T_g$ is noticed in differential scanning calorimetry analysis for 5 wt. % clay combinations. The heat properties of the clay-filled composite are perceived to increase gradually and are observed to be the highest at 5 wt. % in comparison to other variants. The presence of clay, in the composite, improves the heat stability of the composites [11–13].

From DSC analysis it was observed that the glass transition temperature (Tg) of blended nanocomposites varied for 0, 1, 2, 3, 4 and 5 wt% clay contents at 430, 432,

433, 433, 432 and 429°C respectively. A 2°C decrease in glass transition temperature is observed for 4 wt% clay filled samples while 4°C decrease in glass transition temperature is observed for 5 wt% clay filled samples when compared with 3 wt% clay samples, where as no change is observed between the 2 and 3 wt. % clay samples. A 4°C decrease in glass transition temperature is observed for 5 wt% clay when compared with 3 wt% clay samples, where as no change is observed between the 2 and 3 wt% clay samples. The reduction in the values of Tg for unsaturated polyester toughened epoxy system may be due to the flexibility imparted by unsaturated polyester to the epoxy matrix [15].

The composite blend specimens show a weight reduction on treatment with solvents. This is understandable as Unsaturated Polyester (UP), that is present in the blend, dissolves in the solvents under study. The composite has good resistance to attack from acids and alkalis. The highly expandable montmorillonite clay has caused maximum swelling in nitric acid in contrast to maximum weight loss in benzene due to its high cation exchange capacity. The enhancement in weight of the composite is due to the penetration of the liquid chemical in the composite; resulting in swelling [14]. The epoxy/UP/clay composite can be used for applications like a) charge storage containers in vehicles (Nanoclay limits the diffusion of solvents into polymer) b) chemical containers (as the composite has good resistance to chemicals) and c) fire proof cables.

## Author details

Chakradhar V.P. Komanduri
Department of Mechanical Engineering, Vardhaman College of Engineering, Hyderabad, Telangana, India

*Address all correspondence to: chakradharkvp@vardhaman.org

## References

[1] Li J, Jian Guo Z. Influence of Polyethylene-Polyamine Surface Treatment of Carbon Nanotube on the TPB and Friction and Wear Behaviour of Thermoplastic Polyimide Composite. Polymer-Plastics Technology and Engineering. 2011; 50:996-999.

[2] Chinnakkannu K. C, Muthukaruppan A, Rajkumar J S, Periyannan G. Thermo-Mechanical Behaviour of Unsaturated Polyester Toughened Epoxy–Clay Hybrid Nano-composites. Journal of Polymer Research. 2007; 14:319-328.

[3] Uday K, Gautam D, Niranjan K. Mesua ferrea L Seed Oil Based Highly Branched Polyester/Epoxy Blends and Their Nanocomposites. Journal of Applied Polymer Science. 2011; 121:1076-1085.

[4] Guo J, Han-Xiong, H, Zhao-Ke C. Rheological Responses and Morphology of Poly-lactide/Linear Low-Density Polyethylene Blends Produced by Different Mixing Types. Polymer-Plastics Technology and Engineering. 2011; 50:1035-1039.

[5] Song Z., Li. Z., Yan-Ying W, Yi. Z., Shi-Bo G., Yu-Bao L. Fabrication of Hydroxyapatite/Ethylene-Vinyl-Acetate/ Polyamide 66 Composite Scaffolds by the Injection-Moulding Method. Polymer- Plastics Technology and Engineering. 2011; 50:1047-1054.

[6] Jha A, Bhowmick A K. Mechanical and Dynamic Mechanical Thermal Properties of Heat and Oil Resistant Thermoplastic Elastomeric Blends of Poly (Butylene Terephthalate) and Acrylate Rubber. Journal of Applied Polymer Science. 2000; 78:1001-1008.

[7] Colakoglu M. Damping and Vibration Analysis of Polyethylene Fiber Composite Under Varied Temperature. Turkish Journal of Engineering and Environmental Sciences. 2006; 30: 351-357.

[8] Vijaya Kumar K R, Sundareswaran V. Mechanical and Damping Properties of Epoxy Cyanate Matrix Composite under Varied Temperatures. Journal of Engineering and Applied Sciences. 2010; 5:106-111.

[9] Alam N, Asnani N T. Vibration and Damping Analysis of Fibre Reinforced Composite Material Cylindrical Shell. Journal of Composite Materials. 1987; 21:348-361.

[10] Erian A. Armanios, Chandra R, Singh S, Gupta K. Experimental Evaluation of Damping of Fiber-Reinforced Composites. Journal of Composite Technology and Research. 2003; 25:1-12.

[11] Chow Wen Shyang. Tensile and Thermal Properties of Poly (butylene Terephthalate)/Organo-Montmorillonite Nanocomposites. Malaysian Polymer Journal. 2008; 3(1):1-13.

[12] Bakare I.O., Okieimen F.E., Pavithran C., Abdul Khalil H.P.S., Brahmakumar M. Mechanical and thermal properties of sisal fiber-reinforced rubber seed oil-based polyurethane composites. Materials & Design. 2010; 31(9):4274-4280.

[13] Jayaramudu J, Jagadeesh D, Varada Rajulu A, Guduri BR. Tensile and Thermal Parameters of Natural Fabrics and Polymer Coating Effect on Hildegardia Fabric. Journal of Reinforced Plastics and Composites. 2009; 29(11):1664-1668.

[14] Varada Rajulu A, Babu Rao G and Lakshminarayana Reddy R. Chemical resistance and tensile properties of

epoxy/polycarbonate blend coated bamboo fibers. Journal of Reinforced Plastics and Composites. 2001; 20: 50-56.

[15] Suneel Bandi, David Schiraldi A. Glass transition behaviour of clay aerogel/poly(vinyl alcohol) composites, M*acromolecules. 2006;* 39(19):6537-6545.

Section 4

# Industrial Application

Chapter 4

# Exploitation of Bentonite for Wastewater Treatment

*Kali Abderrahim, Loulidi Ilyasse, Amar Abdelouahed, Boukhlifi Fatima, Hadey Chaimaa, Jabri Maria and Mbarka Ouchabi*

## Abstract

Bentonite is a clay with interesting surface properties (affinity for water, adsorption capacity for electro-positive compounds ... .). The characteristics and clarifying properties of bentonite from various companies are the subject of numerous studies. The present work focuses on the study of the efficiency of bentonite and modified bentonite to purify aqueous solutions containing organic pollutants such as phenol. First, before starting the adsorption study, a physical–chemical characterization of the clay by FTIR, BET and XRD techniques was undertaken. The specific surface of the bentonite is calculated by BET. Then, the study of isotherms and kinetics of phenol adsorption on commercial BTC showed that this pollutant can be removed from liquid effluents with a significant percentage. Langmuir and Freundlich models were applied. Finally, the kinetic study performed by UV–Visible was reproduced by FTIR spectroscopy.

**Keywords:** bentonite, phenol, adsorption, FTIR, UV–Visible

## 1. Introduction

Bentonite is characterized by high adsorption, ion exchange, and swelling capacity, as well as by specific rheological properties (thixotropy). It, therefore, has wide applications, ever more numerous and in different fields (drilling, foundry, ceramics, painting, pharmacy, bleaching earth, ... ). Bentonites play a significant role in a diverse range of environmental problems and their applications are steadily increasing. Among the few fields of application, bentonite can be used for the purification of gases [1], the elimination of radioactive elements [2, 3], the elimination of pesticides [4], and the elimination of phenol. This pollutant was removed by variety of clays and modified clay as the bentonite [5], the clinoptilolite [6], the composite of silica [7], the silica/hydrotalcite [8], the natural clay [9], the Montmorillonite, Clinoptilolite and Hydrotalcite [10], the hectorite [11], the untreated coffee wastes [12], the lignite [13], the zeolite X/activated carbon [14], the shells of eggs [15, 16] and the chitin/chitosan [16–18].

Zohra Dali and all [5] have studied the adsorption of phenol on two types of clays: sediment after chemical activation with ammonium chloride and bentonite which has undergone activation by sulfuric acid. They concluded that the acidified bentonite exhibits more affinity towards phenol with a limited adsorption capacity equal to 32.23 mg/g. Myroslav and al [6] studied the kinetics of phenol adsorption on clinoptilolite modified by hexa decyl trimethyl ammonium (HDTAM). They have shown that phenol can be fixed by this clay with a percentage of (85–90%). The adsorption process is fast. All phenol were fixed after one hour. The adsorption isotherm is described by the Langmuir model.

By studying the interposed action of two kinds of clay (bentonite and kaolinite) by both the surfactant hexa decyl trimethyl ammonium (HDTAM) and the phenyl methyl ammonium bromide (PTMA), Uday and all [19], have shown that the kinetics of adsorption on both clays follows the model of the pseudo-second-order and the adsorption isotherm following the model Freundlich and Langmuir. The parameters thermodynamics shows that adsorption was exothermic and spontaneous. In another research, Jin and al [20], chose to intercalate a Na-montmorillonite with dihydroxy ethyl methyl ammonium bromide (ODEM). The results obtained showed that the adsorption kinetic of phenol by (Na-Mt) modified was pseudo-second-order. The adsorption isotherm described by the Langmuir model gave an adsorption capacity of (384.61 mg/g at 308 K). The parameters thermodynamics indicate that the adsorption process was exothermic and spontaneous. Richards and all [21], for their part, have studied the phenol adsorption kinetics on two clay (basaltic clay and bentonite) changes organically by hexa decyl trimethyl ammonium (HDTAM) and phenyl trimethyl ammonium. The results obtained shows that the adsorption kinetics of (HDTAM) -basaltic and (TMPA) -Bentonite was pseudo-second-order. For its part, Jianfeng and all [22] have studied the adsorption kinetics of phenol and ap-nitro-phenol and β - naphthol on organobentonite (Bentonite-TAB). They showed that phenol can be fixed by this organobentonite after 25 min at 69% and 92% for p-nitrophenol and 99% for β - naphthol. The treatment of phenol by adsorption flocculation using an organobentonite was studied by Yun Hwei [23]. The process consists of dispersing the bentonite in water by adding a cationic surfactant that has a short chain (BTMA). The result showed that BTMA bentonite had a high affinity for phenol and 90% of phenol was removed and 100% bentonite was recovered by the process of adsorption flocculation. The study by Basri and all [24] on bentonite modified by the surfactant cetyl trimethyl ammonium bromide (CTAB) as the adsorbent of phenol taking into account several parameters such as the pH of the solution, the contact time. At the initial concentration and temperature, we showed that the adsorption is maximal at pH = 9. The equilibrium was reached after 1 hour with a kinetic adsorption second order. The calculation of the thermodynamic parameter ($\Delta G°$), enthalpy ($\Delta H°$) and entropy ($\Delta S°$) has shown that the adsorption on the organobentonite is possible spontaneous and exothermic in the range of temperature 0-40C °. The adsorption of phenol and 2-chlorophenol and 2,4,6 trichlorophenol by organo-clays (Na-montmorillonite) modified by transition metal complexes was undertaken by Boufatit and al [25]. The results showed that the adsorption capacity of organophilic clay is not obeyed in a logical order, but depends on the nature of the complex. The work carried out by SYuening and all [26] allowed the study of the elimination of phenol using Na-montmorillonite modified with two new Gemini surfactants containing hydroxyl groups, 1,3-bis (hexa decyl dimethyl ammonio) -2-hydroxy-dichloride (BHHP) and1,3-bis (octyl dimethyl ammonio) -2-hydroxy-dichloride (BOHP) as adsorbent. The effects of contact time, pH, temperature, and concentration of adsorbate on the adsorption performance of phenol by the Na - modified montmorillonite were examined by tests in batch. The results have shown that the kinetics follows the pseudo-second-order model, and the adsorption equilibrium data are described by the Langmuir model. The thermodynamic study has shown that the adsorption of phenol is a spontaneous and exothermic process.

The present chapter is interested in the study of the elimination of phenol as a very toxic pollutant using commercial bentonite (BTC) and treated bentonite. The treatment of the clay was carried out with acid solutions of different concentrations.

Before starting the pollution treatment, physicochemical characterization of the bentonite was undertaken. The elimination of the pollutant was carried out by studying kinetic and isothermal adsorption process.

## 2. Materials and methods

### 2.1 Presentation of the phenol used

In the form of white crystals at ambient temperature, it is of GMBH origin. It has a purity of 99.5%. Phenol solutions were prepared with distilled laboratory water. The main characteristics of phenol are grouped in **Table 1**.

Phenol is weakly acidic and transforms in a basic medium into a phenolate anion according to the reaction:

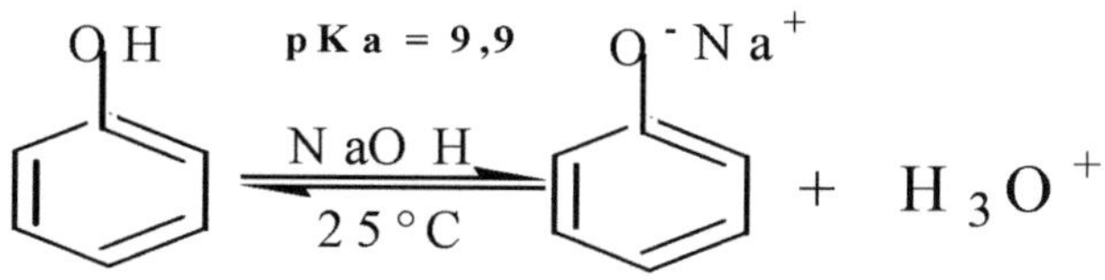


Phenol is very toxic: VA = 19 $mg.m^{-3}$.

### 2.2 Procedure

The various phenol adsorption experiments on bentonite are carried out at room temperature. A fixed mass $m_0$ = 0.1 g of bentonite is contacted with the aqueous solution of phenol with an initial concentration of $C_0 = 2.10^{-3}$ M. The mixture is then stirred at 500 rpm for an adsorption time t (tads). For each experiment, the solution pH is adjusted, as necessary, by the addition of HCl (0.1 N) or NaOH (0.1 N).

After the time required for adsorption (tads), the solutions are filtered through a 0.45 μm microporous membrane, and then the filtrates obtained are analyzed by UV/Visible spectrophotometry and the bentonite by IRTF spectrophotometry.

Commercial bentonite (BTC) was treated with HCl 3 N (BTC3N) and HCl 12 N (BT12N) overnight. The objective of these treatments is to determine the effect of acid on the capacity of adsorption by BTC. The study of the adsorption kinetics of phenol and its isotherms was carried out by UV / visible and IRTF.

### 2.3 Spectroscopic techniques for characterization and analysis

#### *2.3.1 IRTF and UV/visible spectrophotometry*

IR and UV / Visible spectrophotometry are two analysis techniques, based on the principle of absorption by the sample of a beam in IR [400–4000] $cm^{-1}$ or UV / Visible [200–800] nm range.

| TF (temperature of melting in °C) | TE (temperature of boiling in °C) | $S/H_2O$ ($gL^{-1}$) (20 °C) | μ (D) | pKa (acidity constant) |
|---|---|---|---|---|
| 41 | 181 | 93 | 1.59 (pH → OH) | 9.9 |

**Table 1.**
*Characteristics of the phenol used.*

- **Spectrophotometer IR** is Fourier transform (FTIR) of the JASCO 4000, provided with a sensor (TGS) and a source ceramic, separated by an optical system using an interferometer Michelson. The operating conditions for analysis are as follows: - Resolution: 2 or 4 $cm^{-1}$ - Initial beam intensity: 22000 - Analysis range: 400–4000 $cm^{-1}$. The solid samples to be analyzed are diluted to 4% by weight of the material in KBr, then finely ground. A mass of 50 mg (48 mg of KBr + 2 mg of BT) was placed in a mold and then subjected to a pressure of 6 tones.

- **UV / visible spectrophotometer**, with a double beam allows the recording of spectra between 200 nm - 800 nm. This technique is generally used for quantitative analysis. We used it for the determination of the quantities of residual phenol, not adsorbed on the bentonite.

#### *2.3.2 SEM/EDX scanning electron microscopy*

Scanning electron microscopy is a spectroscopic technique, based on the principle of electron-matter interaction. An electron beam scans the surface of the powdered solid, deposited on a sample holder, which in response re-emits certain particles (electrons). Different detectors make it possible to analyze these particles and fully reconstruct the image or cartography of the surface of the solid analyzed. Combined with the EDX technique, it provides access to the chemical composition of the solid. This technique is used for the characterization of the texture of bentonite and its chemical composition.

## 3. Results and discussions

### 3.1 Physical-chemical characterization of raw and treated bentonite

Commercial bentonite (BTC) in the form of a very fine white powder was purchased from Rhône Poulenc. Its origin is a volcanic clay rock, strongly colloidal. It is identical to montmorillonite of the smectite group, consisting mainly of alumina silicate. Its crystalline structure is in the form of a sheet of alumina octahedron, placed between two sheets of silica tetrahedron. The percentage of silica is 50 to 60% and that of alumina is between 15 and 20%, the rest is in the form of metal oxides (Fe, Mg, Ti) and oxides of alkaline elements and alkaline earth (Na, K, and Ca) [3]. The amount of interstitial $H_2O$ and the nature of the exchangeable cations present in the interleaf space determine the Physico-chemical properties of bentonite. These essential properties enhance bentonite in its field of use and application.

Before performing the phenol removal experiments, the commercial bentonite BTC and treated with 3 N and 12 N HCl acid were characterized by the techniques of BET, IRTF, and SEM / EDX to evaluate the surface and active functions of clay.

#### *3.1.1 Characterization by the BET method*

The specific surface of the bentonite was evaluated at 19 $m^2/g$ by the BET method using an automatic device of the ASAP type from the European Institute of Membranes (IEM) in Montpellier. **Figure 1** shows the adsorption/desorption isotherm of $N_2$ at −196° C on the solid and in the **Table 2** the parameters characterizing its texture. The isotherm thus obtained is of type IV, according to the IUPAC classification, exhibiting a hysteresis characteristic of mesoporous solids of type $H_3$.

For BTC 3 N and BTC 12 N, the specific surface areas are equal to 96 $m^2/g$, greater than the surface area of BTC, indicating an increase in nitrogen adsorption sites.

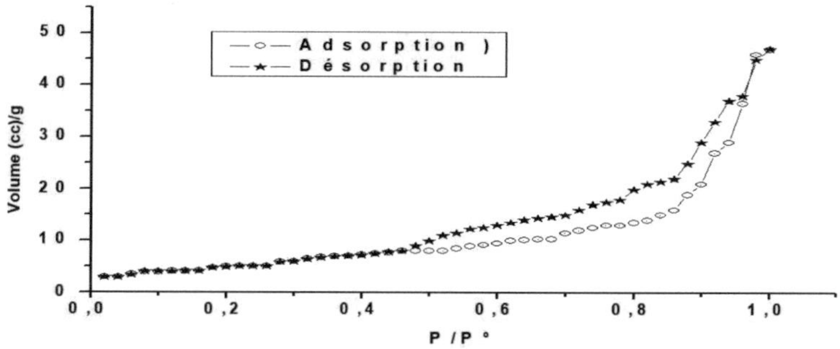


**Figure 1.**
*Adsorption/desorption isotherm of $N_2$ on bentonite.*

| Methods | Values | Units |
|---|---|---|
| Point B method | 18.5351 | Specific area ($m^2$ / g) |
| BET | 19.4377 | |
| BJH adsorption | 20.2408 | |
| BJH desorption | 28.9052 | |
| BET | 0.058366 | Volume ($cm^3$ / g) |
| BJH adsorption | 0.070829 | |
| BJH desorption | 0.072696 | |
| BET | 120.1091 | Diameter (Å) |
| BJH adsorption | 139.9721 | |
| BJH desorption | 100.5997 | |
| *With BJH indicates pore size.* | | |

**Table 2.**
*Textural characteristics of bentonite.*

For many technical applications, raw bentonites must be subjected to a preparation adapted to the requirements of their use (activation). Activation with acids such as hydrochloric acid increases porosity by the peripheral dissolution of Smectites. This results in a product with a high adsorption capacity. They are used for clarification or protein stabilization operations in musts and wines [27].

### *3.1.2 Characterization by SEM/EDX*

Scanning electron microscopy images, taken at different scales and two different locations, under the energy of 30 keV, are shown in **Figure 2** for BTC, BTC 3 N, and BTC 12 N. They show that BTC is not treated with HCl, which has a mesoporous sheet structure and variable particle size. Their values are in perfect agreement with the results of BET and BJH. On the other hand, for treated BTC, the SEM images show an increase in the number of pores and a decrease in their size with the concentration of HCl. The surface of BTC 12 N becomes very compact indicating the formation of several micro-pores.

### *3.1.3 Characterization by IRTF*

The IR spectrum of **Figure 3** shows the absorption bands of commercial bentonite before its contact with phenol. The bands thus observed are characteristic of

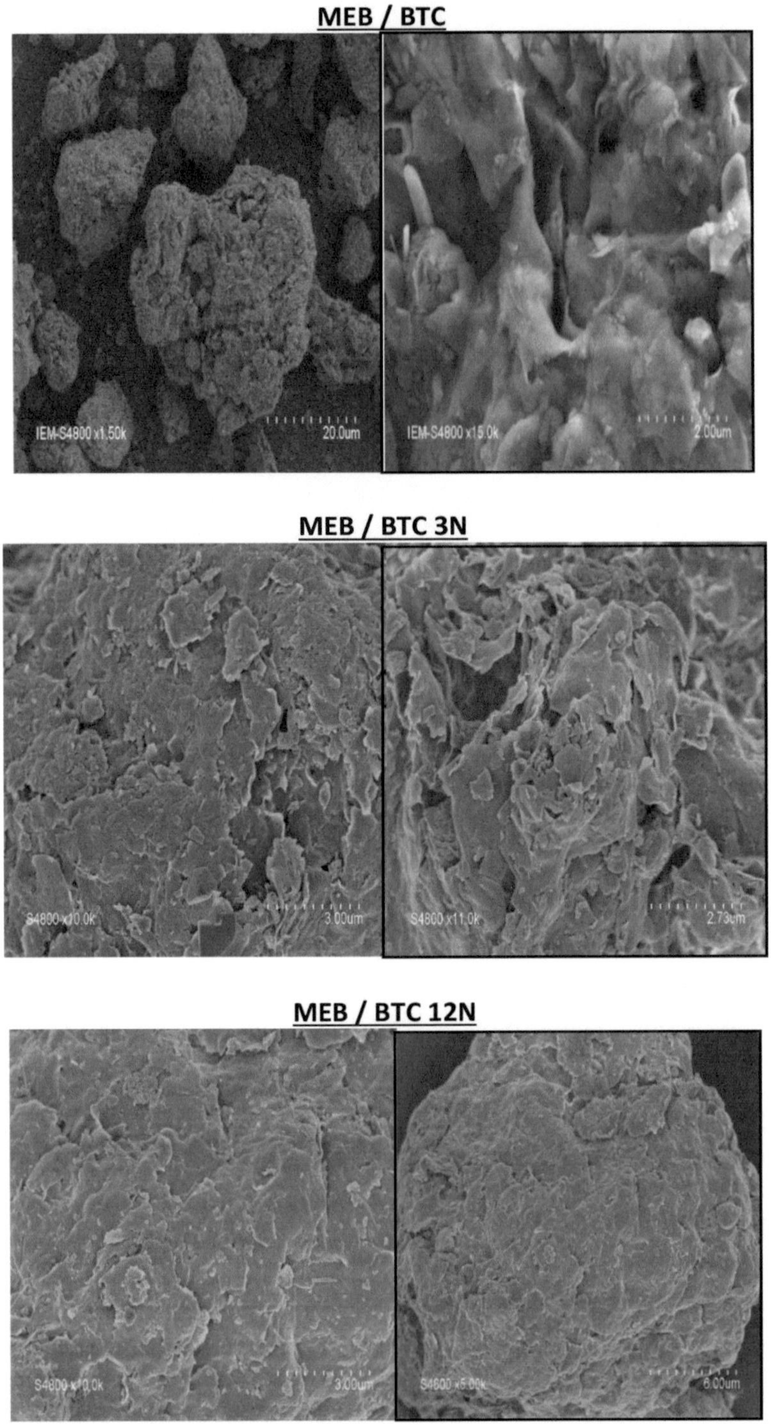


**Figure 2.**
*Microscopic images of BTC, BTC 3 N, and BTC 12 N.*

the structure of clays belonging to the smectite class, such as montmorillonite and bentonite. **Table 3** shows the different frequencies and their attributions to the different vibration modes [28, 29].

The comparison between raw and treated bentonite has been illustrated in **Figure 4**. This figure represents a comparison of the IR spectra of the three samples:

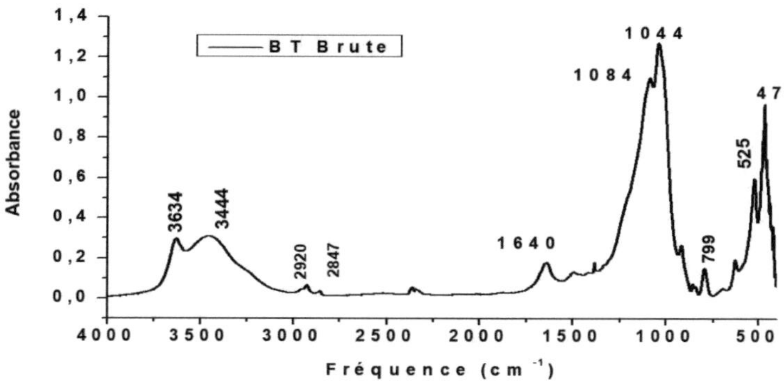


**Figure 3.**
*IRTF spectrum of commercial bentonite.*

| $\nu$ (cm $^{-1}$) | Vibrations | Vibration type |
|---|---|---|
| 477 | Si-O-Si | vibration of distortion out of plane |
| 525 | Si-O-Al or Mg, Fe | vibration of distortion out of plane |
| 630 | Al-O and Si-O | vibration of distortion out of plane |
| 795 | Quartz | Si-O vibration |
| 845 | Al-Mg-OH | Vibration of valence |
| 875 | Al-Fe-OH | The vibration of valence (Al; Mg) |
| 917 | Al- AlO -H | The vibration of valence OH (Fe, Mg) |
| 1035 | Si-O- Si tetrahedron | Asymmetric vibration in the plane |
| 1084 | Al-OH | Vibration in the plane |
| 1494 | OH ($H_2O$) | $H_2O$ deformation vibration |
| 1640 | $CH_2$: alkanes | The vibration of deformation CH bond of |
| 2920 2847 | $H_2O$ | H bond or of combination (2x1640) |
| 3252 | Interfoliar $H_2O$ | Asymmetric and symmetrical vibration |
| 3444 | Si-OH; Al-Al-OH | Free OH elongation vibration |
| 3634 | | |

**Table 3.**
*IR bands of commercial bentonite.*

BTC, BTC 3 N, and BTC 12 N, before adsorption of the phenol. It can be observed that the acid treatment of BTC with 3 N HCl leads to a slight increase in the bands initially identified on BTC in the [3700–1600] cm$^{-1}$ domain compared to BTC 12 N and BTC 6 N. However, in the [400–1500] cm$^{-1}$ domain, this evolution is not observed since the intensities of the bands decrease in the following order: BTC 3 N > BTC 6 N > BTC 12 N, the treatment of alumino-silicates with concentrated acids generates a change in the Si / Al ratio, the major element in the composition of bentonite and on the occasion of new MO bonds (M: Al, Si, or Mg) of the same type as those characterized for BTC can appear and contribute to the increase of these intensities. These bonds (MO) can combine in an acidic medium with the $H^+$ protons to form hydroxyl groups, which would explain the increase in the band at 3630 cm$^{-1}$ (isolated or terminal OH). The increase in the band at 3435 cm$^{-1}$, linked to the interfoliar $H_2O$, may be due to its adsorption and that at 3252 cm$^{-1}$ to the strengthening of the hydrogen bond between water and OH groups [28].

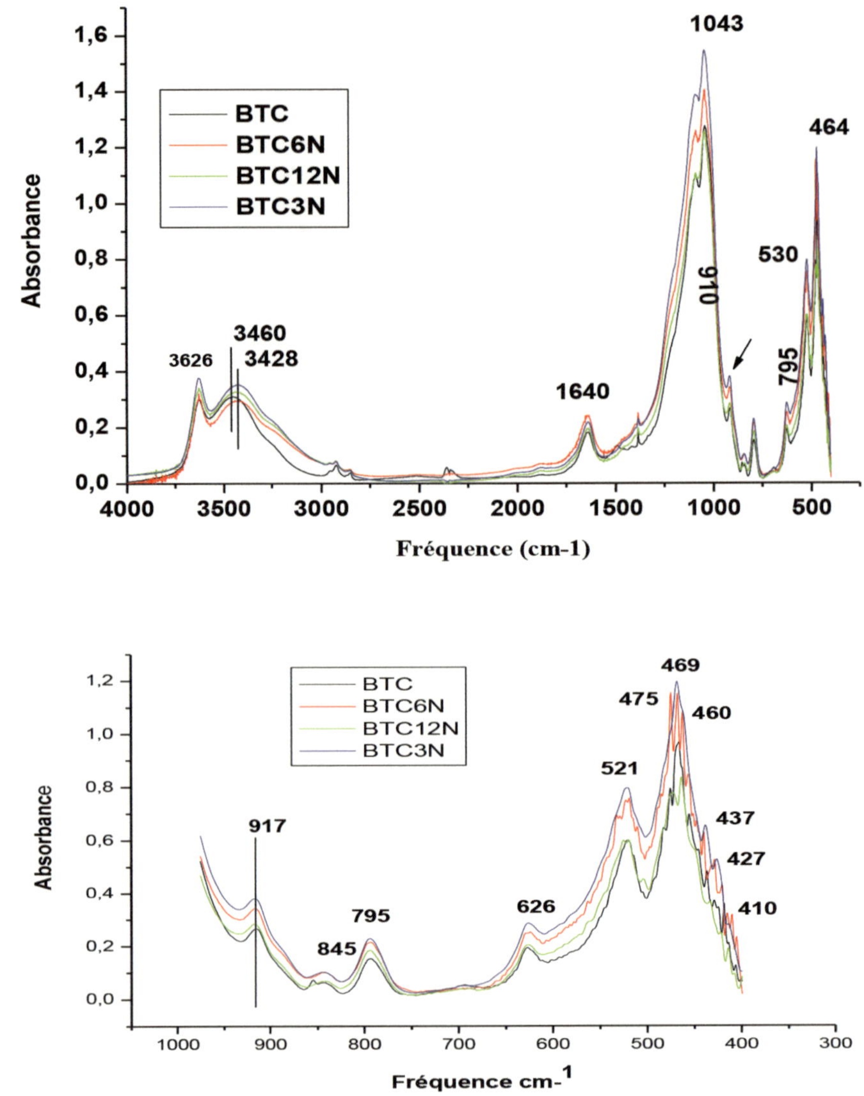


**Figure 4.**
*IR spectra of BTC, BTC 3 N, and BTC 12 N.*

On the other hand, these results are in good agreement with the increase in the specific surface which goes from 19 $m^2$ /g for BTC to 96 $m^2$/g for BTC 3 N and BTC 12 N.

## 3.2 Kinetics study of phenol adsorption on bentonite

### *3.2.1 Study by UV/visible spectrophotometry*

#### *3.2.1.1 Adsorption kinetics of phenol on BTC*

The phenol adsorption kinetics were studied on commercial bentonite (BTC) and bentonite had undergone treatment for 12 hours with 3 N HCL (BT3N) and 12 N HCl (BT12N). To determine the time necessary for obtaining the adsorption equilibrium, phenol adsorption experiments ($C_0$ = $2.10^{-3}$ M) were carried out for different contact times with a mass m = 0.1 g of commercial and treated bentonite.

It can be observed from the UV spectra of the residual phenol in **Figure 5**, the presence of two peaks of phenol, successively at max = 210 nm and 270 nm. Their position does not change concerning that of the pure phenol, which indicates that there is no reaction in solution between the phenol and the species released by the bentonite. The presence of two shoulders at λmax = 230 nm and λmax = 287 nm, is linked to the formation of the phenolate anion, which is favored in a basic medium. An increase in the pH value of the solutions from 8.9 to the value 10, which is equal to the pKa of phenol, was recorded for the stirring times. This increase in pH is explained by the release of the basic species by the BTC, in particular the $Na^+$ cations.

On the other hand, the intensities of the peaks gradually decrease with the contact time, which indicates that the adsorption of phenol on BTC is favored in a basic medium.

By taking into account the deconvolution of the spectra, the exact intensities of the two peaks can be determined. The residual concentration of phenol Cr is in this case represented by the maximum absorbance of the first peak (Ar) in the calibration curve. This will allow the determination of the amount of phenol adsorbed Cards for different contact times using (Eq. (1)), namely:

$$C_{ads} = \frac{C_0 - C_r}{m_{BT}}.M.Vsol = \frac{(A_0 - A_r)}{m_{BT}}.M(p\check{h}enol).Vsol \quad (mg/g) \qquad (1)$$

where $A_0$ is the maximum absorbance of the peak at 270 nm for a concentration of phenol alone ($C_0 = 2.10^{-3}$ M; $A_0 = 2.71$). The use of the above spectra leads to the phenol adsorption kinetics curve given in **Figure 6**. This curve shows that the adsorbed quantity of phenol rises rapidly as a function of time (t < 5 h) and then keeps a practically constant value for longer times. The saturation value of the

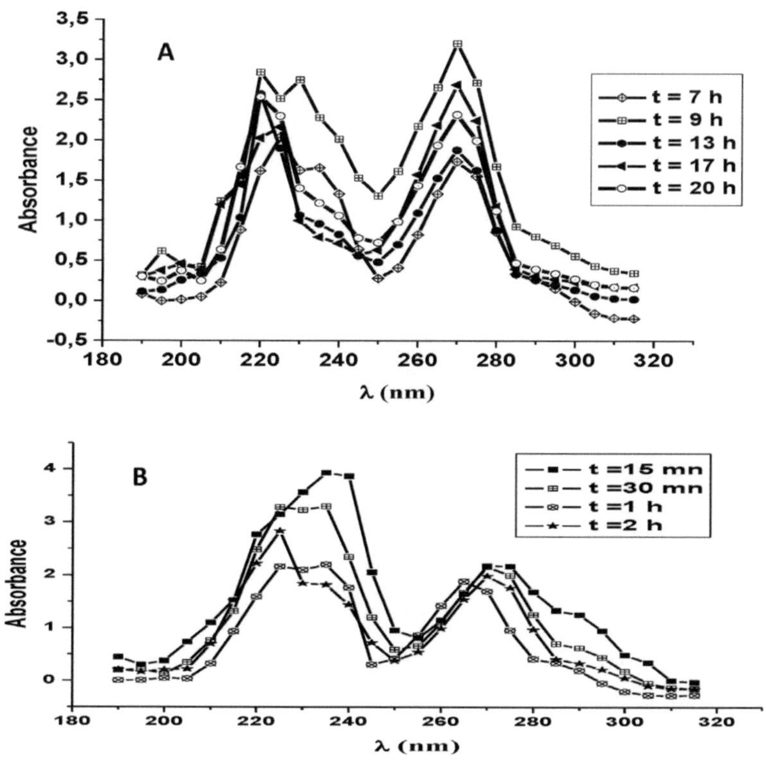


**Figure 5.**
*(A and B) Evolution of the residual concentration of phenol after different adsorption times on bentonite.*

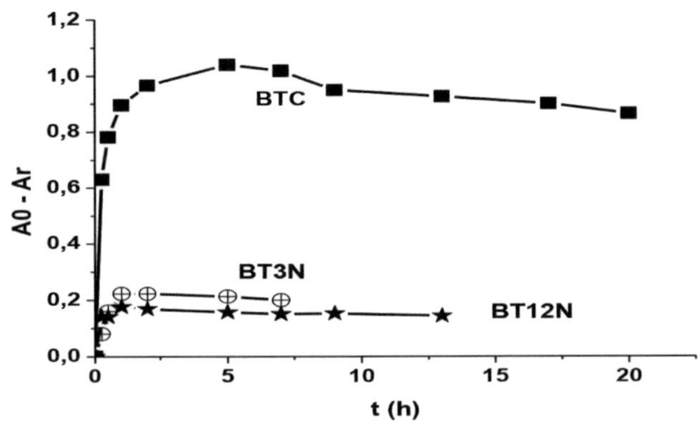


**Figure 6.**
*Adsorption kinetics of phenol (Co = 2.10$^{-3}$ M) on BTC / BTC3N / BTC12N.*

surface corresponds to a concentration of the adsorbed phenol Cad = 8.10$^{-4}$ M, which indicates, for $C_0$ = 2.10$^{-3}$ M, a rate of elimination of the phenol of 40%. The linear rise observed may indicate the adsorption of phenol on equivalent sites, and the plateau obtained reflects a limited number of adsorption sites.

The same evolution is observed for the adsorption of phenol on BT3N and BT12N with a significant drop in the quantity of phenol adsorbed at saturation. The acid treatment of BTC resulted in a significant loss of adsorption sites (**Figure 6**), following the results of SEM analysis, which showed a significant decrease in the pore size of the solids treated, which prevents access to the solids adsorption sites.

*3.2.1.2 Kinetic models*

To elucidate the process of adsorption of phenol on BTC, several kinetic models have been proposed by Lagergren et al. [30], Ho et al. [31] and Boukhlifi et al. [32, 33] by considering pseudo-first–first-order and pseudo-second-order kinetics. The results obtained in this work can be subject to equations resulting from these models [30–33].

*3.2.1.2.1 Pseudo first-order kinetics*

Either phenol adsorption reaction on BTC:

$$\mathrm{Ph} + * \leftrightarrow \mathrm{Ph}*\,(\mathrm{Ph} = \text{phénol})$$
$$\mathrm{C_r}\ (\mathrm{S_0} - \mathrm{S})\ \mathrm{S}$$

Cr: residual concentration of phenol, S and $S_0$ are, respectively, the area occupied by phenol and the total area of BTC.

The differential equation that results from the adsorption reaction is as follows (Eq. (2)):

$$\frac{dS}{dt} = K_1.C_r.(S_0 - S) \Rightarrow \frac{dq_t}{q_e.dt} = K_1'.\left(1\text{-}\frac{q_t}{q_e}\right) \quad (2)$$

where qt, in (mg / g) is the amount of phenol adsorbed at t; QE, in (mg / g) is the quantity of phenol at adsorption equilibrium and K '(min$^{-1}$), the pseudo-rate constant.

Using the initial conditions (at, t = 0; qt = 0) and (at, t; qt), its rearrangement after integration gives (Eq. (3)):

$$\ln\left(q_e - q_t\right) = \ln q_e - k_1'.t \tag{3}$$

The plot of ln (qe - qt) as a function of a (t), given in **Figure 7**, does not lead to a straight line as provided by the model equation. The curve obtained from the experimental data seems rather formed of two portions of straight lines. The first before t = 5 h (**Figure 7**) of negative slope can conform with the kinetic model of pseudo-first-order where the adsorption of phenol occurs in a very fast way (kinetic figure). However, the second portion, with a positive slope (**Figure 7**), is in contradiction with the equation of the proposed model. The linear regression coefficients $R^2$ of the two lines are not close to 1, which makes it possible to conclude that the kinetics of adsorption of phenol on BTC cannot be described by a pseudo-first-order speed. In addition, the calculated value of qe (qe = 47 mg / g) is much higher than the experimental value (≈14.5 mg / g).

*3.2.1.2.2 Pseudo second order kinetics*

The same treatment can be carried out for the pseudo-second-order kinetics (Eq. (4)), namely:

$$\frac{dq_t}{dt} = k_2'.\left(q_e - q_t\right)^2 \Rightarrow \frac{t}{q_t} = \frac{1}{K_2'.q2_e} + \frac{1}{q_e}.t \tag{4}$$

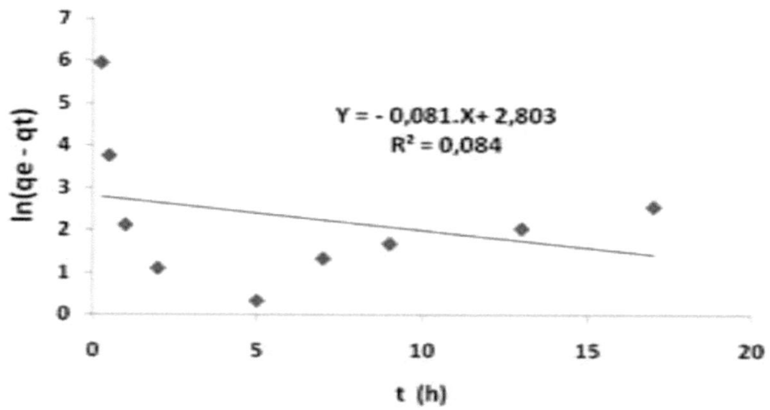


**Figure 7.**
*Kinetics modeling according to the pseudo-first-order model ln ($q_e$ - $q_t$) as a function of a (t).*

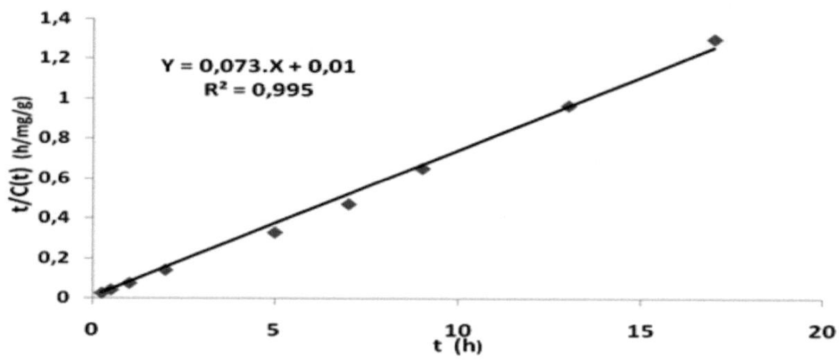


**Figure 8.**
*Kinetics modeling according to the pseudo second order model t/qt = f (t).*

The application of this equation to the experimental result of **Figure 8**, leads in this case to plotting t / qt = f (t)) to the right of **Figure 8**, with a linear regression coefficient ($R^2$ = 0.995) very close to 1. The parameters K′2 and qe, determined from the slope (1 / qe) and the y-intercept (1 / $K'_2$.$qe^2$) of the equation of the line, are respectively equal to 0.532 g / (mg.h) and 13.7 mg / g. The value of qe obtained is in perfect agreement with the experimental value (≈14.5 mg / g), which indicates that the adsorption of phenol on bentonite obeys pseudo-second-order kinetics for the entire time range studied.

*3.2.1.2.3 Intra-particule diffusion model*

The diffusion phenomenon (adsorbate ⇒adsorbent) also plays a determining role in the adsorption kinetics. The equation resulting from the treatment of the intra-particle diffusion model is given by the following relation (Eq. (5)) [34–36]:

$$q_t = K_d.\sqrt{t} + C \qquad (5)$$

where Kd is the intraparticle diffusion constant and C a constant. In the case where C = 0, the intra-particulate diffusion step is the step that controls the adsorption process, if C is nonzero this step is not limiting.

The plot $q_t = f(\sqrt{t})$ leads to the curve of **Figure 9**, made up of two phases: the first increasing for low contact times (t < 5 h) and the following decreasing for longer times. The line corresponding to the first phase shows that C = 7.276 ≠ 0 and Kd = 5.087 mg / (g.h1 / 2). This result is not adapted to the hypothesis of the model on C and consequently, the diffusion process is not limiting.

*3.2.2 Study of phenol adsorption on raw and treated bentonite by IR spectrophotometry*

The masses of the solids resulting from the BTC/phenol contact were analyzed by IRTF. **Figure 10** shows the evolution of the spectra obtained after t = 1 hour and 17 hours of contact. The overall analysis of these spectra shows a clear decrease in the absorption bands of BTC, located at 477, 525, 1035, 1084, and 3634 $cm^{-1}$ and related to the vibrations of the M-O bonds where M can represent the atoms of Si, Al, or Fe (**Table 2**). The decrease in these bands can be explained by their direct

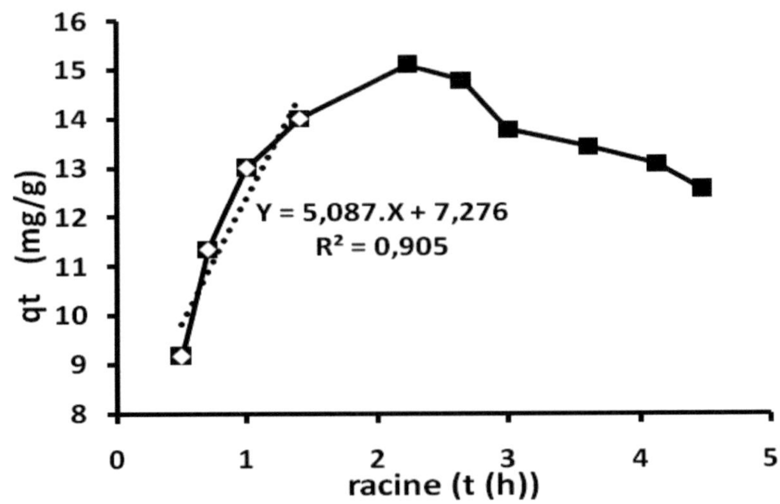


**Figure 9.**
*Intra-particle diffusion model.*

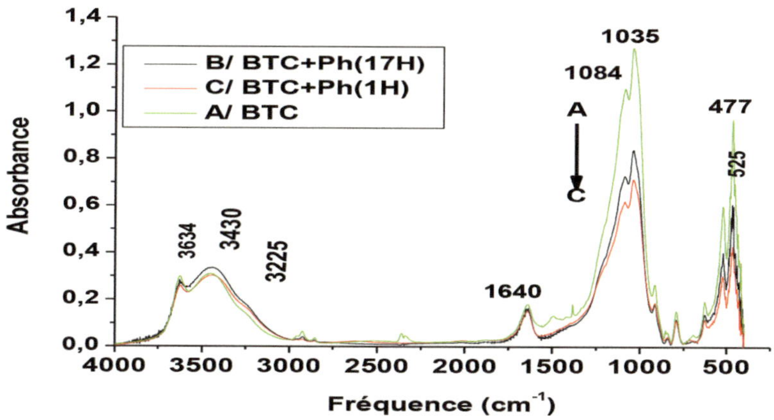


**Figure 10.**
*IR spectra of phenol adsorption on BTC.*

involvement in the adsorption of phenol on different sites of BTC. Their positions, which remain unchanged, meaning that the adsorption of phenol occurs at superficial sites. Indeed, for these two contact times, the intensities of these bands are practically equal, following the quantities of the adsorbed phenol, determined previously (**Figure 6**).

On the other hand, the bands located at 3430 and 1634 $cm^{-1}$, attributed to the OH vibrations of the interfoliar water and the shoulder at 3225 $cm^{-1}$, attributed to the OH group linked by the hydrogen bonds (H ... OH), increase slightly in intensity. For the first band, there is probably the formation of interfoliar water by dehydroxylation and for the last band, the formation of hydrogen bonds during the adsorption of phenol. The formation of $H_2O$ during the adsorption of phenol on clay solids have already been mentioned by several authors according to the following process:

$$C_6H_5\text{-O-}\boxed{H \;+\; H\text{-O}}M \longrightarrow H_2O + C_6H_5\text{-O-M}$$

This proposition is in good agreement with our results which simultaneously show the increase in the $H_2O$ bands and the decrease in the vibration band of the O -H groups at 3634 $cm^{-1}$. Moreover, the mechanism of adsorption of phenol by hydrogen bonding has also been mentioned by SYuening and al [26]. This bond, characterized by the band at 3225 $cm^{-1}$, results from an interaction between phenol and the oxygen atoms of the surface. It appears more marked for t = 17 h than t = 1 h of adsorption, following the intensities of the OH (3636 $cm^{-1}$) and interfoliar $H_2O$ (3430 $cm^{-1}$) vibration bands.

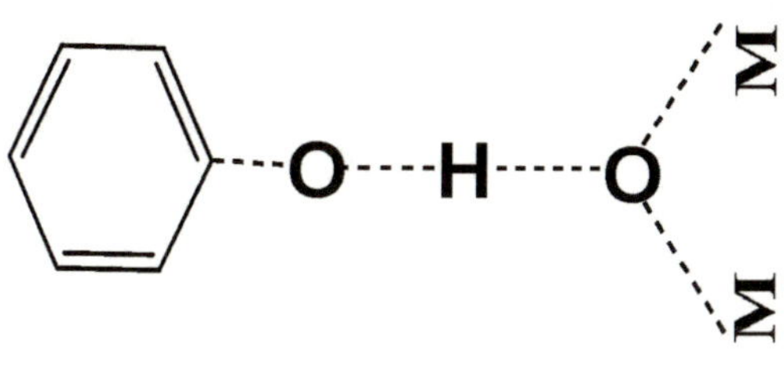


Adsorption of phenol on bentonite treated with 3 N and 12 N HCl was also monitored by IR. The spectra obtained are illustrated in **Figures 11** and **12**. It can be

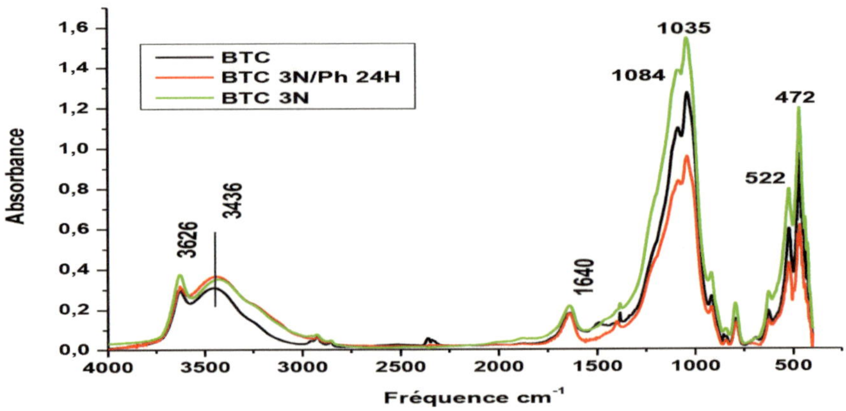


**Figure 11.**
*IR spectra of phenol adsorption on BTC 3 N.*

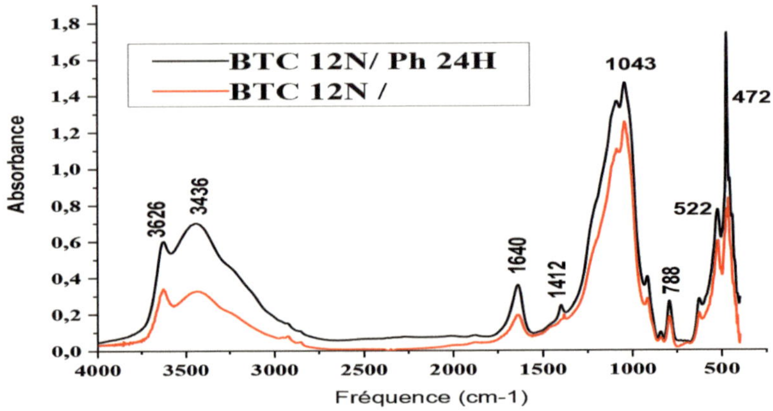


**Figure 12.**
*IR spectra of adsorption of phenol on BTC 12 N.*

noted first that the bands detected on BTC (spectrum A) are also observed at the same positions on BTC 3 N (spectrum B) with an increase in their intensities. This means that the acid treatment leads, in the range [300–1100] $cm^{-1}$, to the creation of MO bonds of the same type as those observed in BTC (Metal: Si, Al, and Fe) probably by breaking the OMO bonds or MOM. This explanation is confirmed on the one hand, by the increase in the intensity of the band to 3626 $cm^{-1}$, often attributed to the vibration of the MOH (free OH) bond and on the other hand, by the increase in the specific surface area which goes from 19 $m^2$ / g for BTC to 96 $m^2$ / g for BTC 3 N. The band at 3436 $cm^{-1}$, linked to the formation of the hydrogen bond increases in intensity under the effect of the acid treatment of BTC.

Furthermore, the adsorption of phenol on BTC 3 N leads to a reduction in the intensities of the characterized bands, except for the bands located at 3436, 3225, and at 1640 $cm^{-1}$. Taking into account the kinetic curve of **Figure 6**, it was found that the amount of phenol adsorbed by BTC 3 N is lower than that adsorbed by BTC.

### 3.3 Isothermal study of the adsorption of phenol

The adsorption isotherm was obtained at T = 25° C, by bringing different initial concentrations of $C_0$ of phenol into contact with BTC for a period t = 5 hours at

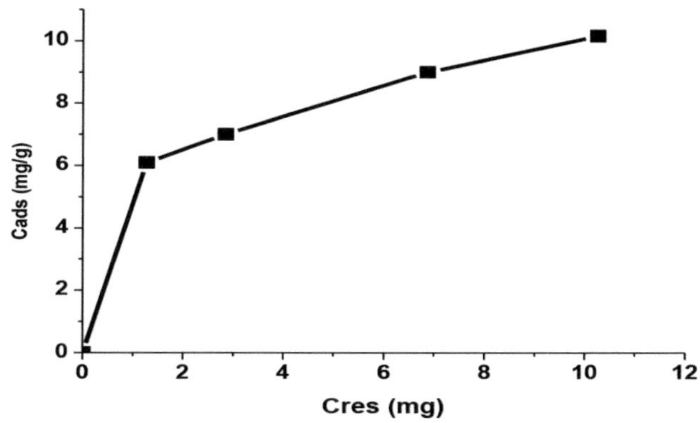


**Figure 13.**
*Phenol adsorption isotherm.*

pH = 6. The study of this isotherm is essential since it makes it possible, from known models, to obtain information on the adsorption parameters characterizing the adsorbate/adsorbent interaction.

The plot of the amount of adsorbed phenol (Cads) as a function of Cr gives the curve of the corresponding isotherm, represented by **Figure 13**. Several models are used to exploit the adsorption isotherms, the most commonly adapted to the adsorption of liquids are the isotherms of Langmuir (homogeneous sites) and Freundlich (heterogeneous sites). The latter is used to determine the parameters giving information on the nature of the interaction (adsorbate/adsorbent).

### *3.3.1 Langmuir model*

This model assumes that the adsorption sites are energetically equivalent and there is no interaction between the adsorbates at two neighboring sites. The equation that governs this model is as follows (Eq. (6)):

$$q_e = \frac{Q_{max}.b.C_e}{1 + b.C_e} \tag{6}$$

Its linearization makes it possible to determine the constants Qmax and b (Eq. (7)), namely:

$$\frac{1}{q_e} = \frac{1}{b.Q_{max}} \cdot \frac{1}{C_e} + \frac{1}{Q_{max}} \tag{7}$$

It can be observed in **Figure 14** that the experimental results do not verify the Langmuir model since the plot of 1 / qe = f (1 / Ce) does not give a straight line. Besides, the theoretical line obtained by linear regression of the experimental points gives a correlation coefficient $R^2$ = 0.938, less than 1. The constants Qmax and b deduced from the ordinate at the origin and the slope of this line are, respectively, equal at Qmax = 10.42 mg / g and b = 1.05 m/g. Another parameter noted RL given by the relation below (Eq. (8)), characterizes a constant of the equilibrium of the Langmuir model. It makes it possible to predict whether such a model is favorable or not according to certain conditions on the values of RL:

$$R_L = \frac{1}{1 + b.C_0} \tag{8}$$

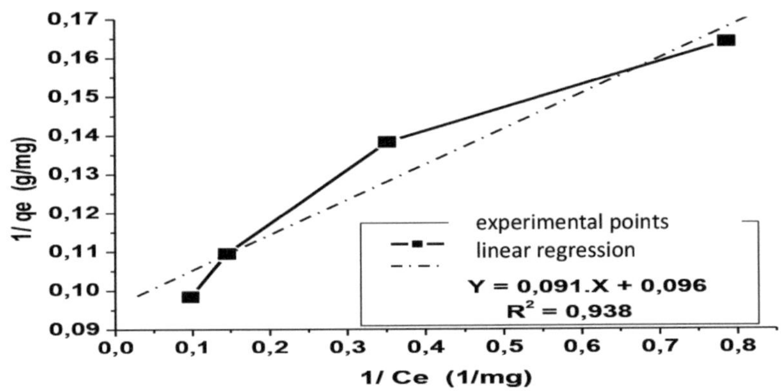


**Figure 14.**
*Langmuir isotherm for the adsorption of phenol on BTC.*

If 0 < RL <1: the isotherm is favorable; if RL = 1: the isotherm is linear and if RL > 1: the isotherm is not favorable and if RL = 0: the isotherm is irreversible.

The calculation of RL is carried out for C0 (max) = 6.10-3 M = 11.28 mg, which gives for RL a value of 0.08. This value is very close to 0 if one takes into account the uncertainties on b and C0, which confirms that the Langmuir model is not favorable.

### *3.3.2 Freundlich model*

This model is often applied in liquid / solid adsorption, it takes as the main assumption a heterogeneity of the solid adsorption sites. The empirical equation established by Freundlich is as follows (Eq. (9)):

$$q_e = K_F.C_e^{1/n} \tag{9}$$

where K, $_F$, and n are the Freundlich constants, related, respectively, to the adsorption capacity and its intensity. Applying this equation in its logarithmic form to the experimental data leads to the values of these constants from the y-intercept and the slope of the following line (Eq. (10)):

$$\ln(q_e) = \ln(K_F) + \frac{1}{n}.\ln(C_e) \tag{10}$$

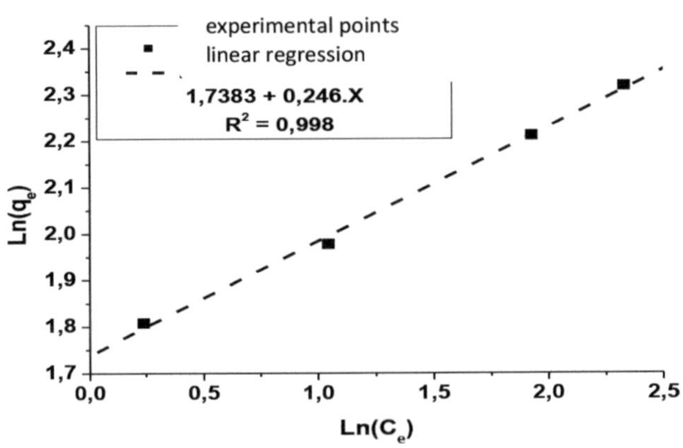


**Figure 15.**
*Freundlich isotherm for the adsorption of phenol on BTC.*

**Figure 15** shows the obtaining of a perfect line with a correlation coefficient very close to 1 ($R^2$ = 0.998), for the initial concentrations used, between $10^{-3}$ M and $6.10^{-3}$ M. The values found for the constants KF and n are, respectively, equal to 5.68 mg / g and 4.06. The low value of (1 / n = 0.246), indicates that the Freundlich model is more suitable for describing the adsorption isotherm of phenol on a BTC surface having very heterogeneous sites, the number of which is low if the value of KF is taken into account.

## 4. Conclusion

The study of the adsorption of phenol on commercial BTC has shown that this pollutant can be eliminated from liquid effluents with a fairly large percentage (40%). This encouraging result allows us to extend this study to real industrial discharges. The treatment of bentonite with concentrated HCl (3 N) and (12 N) did not give satisfactory results, although it led to an increase in the specific surface from 35 $m^2/g$ for bentonite not treated to 96 $m^2/g$ for bentonite treated with acid. This is due to a decrease in the number of phenol adsorbing sites and the decrease in the pore diameters as shown by the analyzes by BET and SEM. The modeling of the experimental results of the adsorption isotherm made it possible to specify the most appropriate kinetic model, the adsorption is fast before two hours. It is the Freundlich model which is favorable for a surface having heterogeneous sites. The adsorption capacity was evaluated at 5.68 mg/g. The mechanism of phenol adsorption by hydrogen bonding has also been proven by spectroscopic analyzes. This bond, characterized by the band at 3225 $cm^{-1}$, results from an interaction between phenol and the free oxygen doublet of the surface.

## Author details

Kali Abderrahim, Loulidi Ilyasse, Amar Abdelouahed, Boukhlifi Fatima*, Hadey Chaimaa, Jabri Maria and Mbarka Ouchabi
Laboratory of Chemistry-Biology Applied to the Environment, Faculty of Sciences, Moulay Ismail University, Meknes, Morocco

*Address all correspondence to: f.boukhlifi@umi.ac.ma

## References

[1] Melnitchenko A, Thompson JG, Volzone C, Ortiga J. Selective gas adsorption by metal exchange d amorphous kaolinite derivatives. Applied Clay Science. 2000;**17**:35-53

[2] Dultz S, Bors J. Organophilic bentonite as adsorbents for radionuclides II. Chemical and Mineralogical Properties of HDPY-montmorillonite. Applied Clay Sciences. 2000;**16**:15-29

[3] Hennig BC, Reich T., Dahn R., Scheidegger AM. Structure of uranium sorption omplexes at montmorillonite edge sites. Radiochem. 2002; Acta 90, 653-657.

[4] Bojemueller E, Nennemann A, Lagaly G. Enhanced pesticide adsorption by thermally modified bentonites. Applied Clay Science. 2001;**18**:277-284

[5] Zohra Dali-Youcefa, Hassiba Bouabdasselema, Nourredine Bettaharb, Élimination des composés organiques par des argiles locales, Comptes Rendus Chimie, 2006;**9**(10):1295-1300.

[6] Sprynskyy M. Tomasz Ligor, Kinetic and equilibrium studies of phenol adsorption by natural and modified forms of the clinoptilolite. Journal of Hazardous. 2009;**169**:847-854

[7] Fuqiang Year Baojiao Gao adsorption mechanism and property of novel composite material PMAA / SiO2 Towards phenol, Chemical Engineering Journal 153 (2009) 108-113.

[8] Yu Fei Tao, Wei Gang Lin, Low-cost and effective phenol and basic dyes trapper derived from the porous silica coated with hydrotalcite gel, Journal of Colloid and Interface Science 358 (2011) 554–561.

[9] A. Amar, I. Loulidi, A Kali, F BoukhlifI,1C. Hadey, M. Jabri, Physicochemical Characterization of Regional Clay: Application to Phenol Adsorption", Applied and Environmental Soil Science Volume 2021, ID 8826063 | https://doi.org/10.1155/2021/8826063

[10] Yapar S, Yilmaz M. Removal of Phenol by Using Montmorillonite, Clinoptilolite and Hydrotalcite. Adsorption. 2005;**10**:287-298

[11] Feng Zhen, Structure and properties of hectorite synthesized by microwave and its application, BioChem-Engineering Department, Jiangsu Food College,Jiangsu Huai'an 223001,China.

[12] Ilyasse Loulidi, Fatima Boukhlifi, Mbarka Ouchabi, Abdelouahed Amar, Maria Jabri, Abderahim Kali, and Chaimaa Hadey, Assessment of Untreated Coffee Wastes for the Removal of Chromium (VI) from Aqueous Medium, International Journal of Chemical Engineering, Volume 2021, 11 pages.

[13] Polata H, MolvabM M. Polatc. Capacity and mechanism of phenol adsorption on lignite, International journal of mineral processing. July 2006; **79**(4):264-273

[14] Wen Ping Cheng, Wei Gao, Xinyu Cui, Jing Hong Ma, Rui Feng Li, Phenol adsorption equilibrium and kinetics on zeolite X/activated carbon composite, Journal of the Taiwan Institute of Chemical Engineers, Volume 62, 2016,

[15] Boukhlifi F., Chraibi S, Alami M, "Evaluation of the adsorption kinetics and equilibrium for the potential removal of phenol using a new biosorbent ", Journal of Environment and Earth Science Vol. 3, No. 7. 2013.

[16] Boukhlifi F, Bencheikh A, Ahlafi H. Characterization and adsorption propriety of chitin toward copper $Cu^{2+}$. Physical and Chemical News. 2011;**58**: 67-72

[17] Boukhlifi F: characterization and treatment of real wastewater from an electroplating company by raw chitin. In: Basheer Al-Naib editor. Book chapter: Recent Advancements in the Metallurgical Engineering and Electrodeposition 2020, Intechopen ISBN 978-1-78984-687-4.

[18] Boukhlifi F: New Approaches and Recent Advances sustainable treatment of heavy metals by adsorption on raw chitin. In: Dr. Mario Alfonso Murillo-Tovar editors, Chapter in book: Trace Metals in the Environment, 2020Intechopen. ISBN 978-1-83880-332-2.

[19] Uday F. Alkaram, Abduljabar A. The removal of phenol from aqueous solutions by adsorption using surfactant-modified bentonite and kaolinite, Journal of Hazardous Materials 169 (2009) 324–332.

[20] Jin S. Hong-fu Lin, Adsorption of phenol from aqueous solutions by organomontmorillonite. Desalination. 2011;**269**:163-169

[21] Richards S, Bouazza A. Phenol adsorption in organo-modified basaltic clay and bentonite. Applied Clay Science. 2007;**37**:133-142

[22] Ma J, Zhu L. Removal of phenols from water accompanied with the synthesis of organobentonite in a one-step process. Chemosphere. 2007;**68**: 1883-1888

[23] Yun-Hwei Shen, Removal of phenol from water by adsorption–flocculation using organobentonite, Water Research 2002;**36**:1107-1114

[24] Hasan Basri Senturk, Duygu Ozdes, Removal of phenol from aqueous solutions by adsorption onto organomodified; Tirebolu bentonite: Equilibrium, kinetic and thermodynamic study, Journal of Hazardous Materials 172 (2009) 353–362.

[25] M. Boufatit, H. Ait -Amar, Development of an Algerian material montmorillonite clay. Adsorption of phenol, 2-dichlorophenol, and from aqueous solutions onto montmorillonite exchanged with transition metal complexes Desalination 206(2007) 394-406.

[26] SYuening Liu, Manglai Gao *, Zheng Gu, Zhongxin Luo, Yage Ye, Laifu Lu, Comparison between the removal of phenol and catechol by modified montmorillonite with two novel hydroxyl-containing Gemini surfactants, Journal of Hazardous Materials 267 (2014) 71– 80.

[27] Boukhlifi F. Study of the retention of metallic micro-pollutants (Pb, Cd, Cu, and Zn) on new biosorbent materials: purification tests of liquid industrial effluents, doctoral thesis. Morocco: Chouaib Doukkali El university Jadida; 2000

[28] Greesh N, Hartmann PC, Cloete V, Sanderson RD. Colloid and Interface Science. 2008;**319**:2-11

[29] Busca G, Montanari T, Bevilacqua M, Finocchio E. Inorganic Solids Colloids and Surfaces A: Physicochem. Eng. Aspects. 2008;**320**:205-212

[30] Lagergren, S, K. Svenska. Vetenskapsad. Handl. 24,1 (1898).

[31] Ho. G. McKay. Sorption of dye from aqueous solution by peat, Chem.Eng J. 70 ((1998115–124.

[32] Boukhlifi F, El Akili C, Moussout H, Benzakour A, Ahlafi H. Treatment of Global Rejection of Electroplating Industry by Raw Chitin. Int. J. Appl. About. Sci. 2013;**8**:13-23

[33] Boukhlifi F, Amar A., Loulidi I, Jabri M, Benzakour, A, purification of the waste water of the galvanoplasty industry by natural adsorbents: current Rinsing, RHAZES: Green and

Applied Chemistry Vol 2, 8, 2018, pp.81~91.

[34] Loulidi I., Boukhlifi F, Ouchabi M., Amar A, Jabri M, Kali, Adsorptive Removal of Chromium (VI) using Walnut Shell, Research Journal of Chemistry and Environment, Vol. 23 (12) December (2019).

[35] Boukhlifi F., Bencheikh A., Study of the competitive adsorption of heavy metals on raw chitin: application to wastewater from the chemical industry. The Water Tribune, N ° 611/3, vol. 55, pp. 37-43, 2001.

[36] Boukhlifi F., Allali M., Bencheikh A., Study of the metallic pollution of wastewater from the chemical industry and test of treatment with crude chitin, Mar. Life, 11 (1-2), pp: 49-56. 2001.

Section 5

# Testing and Evaluation

Chapter 5

# An Introduction to Montmorillonite Purification

*Hakan Ciftci*


## Abstract

Purification of montmorillonite is a process to remove non clay minerals (gangue) such as calcite, feldspar, quartz, opal (C-T), and mica from montmorillonite ore. This is performed to make montmorillonite suitable for use in sensitive applications such as pharmaceutical, cosmetic, food, and advanced materials for nanotechnology. Gangue minerals in raw montmorillonite ores can cause serious health problems when used in pharmaceutics, cosmetic, and food industries and reduce material quality in advanced materials production. Montmorillonite purification can be divided into two main classes as physical and chemical purification. Physical purification processes are based on particle size difference between the gangue and montmorillonite minerals. Purification processes based on gravity separation are ineffective since the specific weights of gangue and montmorillonite minerals are very close to each other. Physical purification process includes sedimentation, centrifugal separation, aero separation, and sieving techniques. Chemical purification of montmorillonite is based on dissolution and so extraction of carbonates, metal hydroxides, organic materials, and silica, respectively, using different leaching techniques.


**Keywords:** montmorillonite, bentonite, purification, beneficiation, enrichment, leaching

## 1. Introduction

Montmorillonite, a type of smectite mineral, is an aqueous aluminum silicate. It has a layered crystalline structure with the general chemical formula of $(Na,Ca)_{0.33}(Al, Mg)_2\ (Si_4O_{10})(OH)_2{}^*nH_2O$. Rocks that mostly contain montmorillonite mineral are also called as bentonite. For this reason, sometimes the term bentonite is used instead of montmorillonite. Montmorillonite clay has some superior characteristics such as high surface area, high swelling capacity, good adsorbent ability, high cation exchange capacity, plasticity, high chemical stability, and good mechanical properties [1–4]. Due to these properties it exhibits, montmorillonite has many uses and research areas, some of which can be listed as pharmaceutical and cosmetic industry, adsorbents, catalysts, drilling mud, and filler in construction industries such as ceramics [2, 5–7].

Raw montmorillonite ores (bentonite) generally contain gangue minerals such as quartz, opal (C-T), mica, calcite, and feldspar, as well as montmorillonite minerals. While raw montmorillonite is mostly used as raw material for drilling mud, cement, and ceramic, it is used in purified form (montmorillonite) in pharmaceutical, cosmetic, food, and advanced materials industries [8]. For example, thanks to the strong adsorption and detoxification effect of montmorillonite on toxic substances, it is widely used in the treatment of esophagitis, gastritis, colitis, and similar diseases [9]. Purified montmorillonite are also used as raw materials for cosmetics due to their decontamination, detoxification, itching, and beautifying properties [3]. Along with these, montmorillonite is widely used as drug delivery system and food protective material, closely related to our life and health [10]. For such uses, it is necessary to separate montmorillonite minerals from gangue minerals (purification) to ensure health and material quality. For this purpose, different methods have been used in many studies on the purification of montmorillonite.

## 2. Purification of montmorillonite

The specific weight of clay minerals and accompanying gangue minerals (quartz, feldspar, calcite, opal, mica, etc.) are very close to each other (2.4–2.7 g/cm$^3$) [11]. For this reason, gravity separation processes based on specific weight difference in order to separate gangue minerals from clay minerals cannot give successful results. The particle size difference between montmorillonite and gangue minerals ensures several physical separation techniques, however. In addition, chemical purification of clay minerals can be summarized as decomposition of carbonates, dissolution of metal hydroxides, oxidation of organic substances, and dissolution of silica. All purification methods can be grouped into two main classes as physical, and chemical purification as shown in **Figure 1**.

### 2.1 Physical purification

Physical purification methods are the most used techniques for the montmorillonite purification. The processes are easy and based on the difference in particle sizes of montmorillonite and gangue minerals. The particle size distribution of montmorillonite clay is between 0.1 μm and 2 μm and it has an average particle size of 0.5 μm [12]. The particle size of most of the gangue minerals is larger in the raw form and generally ranges from 20 μm to 1 cm, however. The physical purification methods according to particle size difference can be listed as sedimentation, centrifugal separation, aero separation, and sieving.

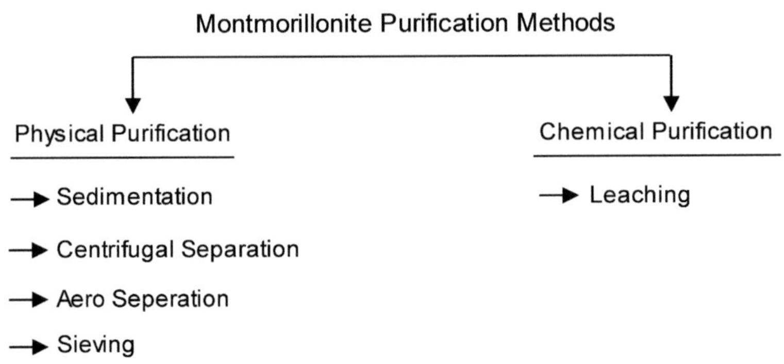


**Figure 1.**
*Purification methods of montmorillonite clay.*

### *2.1.1 Sedimentation*

Clay minerals like montmorillonite are well dispersed in very small sizes in the aqueous media, while gangue minerals remain in coarse sizes. For this reason, the method generally used in montmorillonite purification is sedimentation (settling) in water, depending on the particle size difference. During the sedimentation process, clay minerals dispersed in water in much smaller particle sizes remain suspended, while gangue minerals in coarse sizes settle to the bottom. In this process, suspended clay minerals can be separated from other minerals by decantation. In order to prevent particle/particle interactions during the sedimentation process and thus increase the purification yield and recovery, the solid content in the water should be between 0.5% and 1.0%. Stokes sedimentation equation is generally used to estimate sedimentation time Eq. (1). For an example, a quartz particle with a diameter of 1 μm settles from a height of 0.1 m in approximately 26 h and 25 min. As another example, the dispersion obtained with a sedimentation height of 20 cm and a sedimentation time of 52.5 h contains only particles under 1 μm size [12].

$$t = \frac{18\eta h}{(\rho - \rho o) g d^2} \tag{1}$$

Here; t: sedimentation time (s), η: fluid viscosity (g/cm.s), h: sedimentation height (cm), ρ: solid specific weight ($g/cm^3$), ρo: liquid specific weight ($g/cm^3$), g: gravitational acceleration (9.81 $m/s^2$), d: particle diameter (cm).

An example of laboratory scale sedimentation process for montmorillonite purification is given below, step by step [1].

1. The particle size of raw montmorillonite clay is first reduced to less than 5 mm using a jaw crusher and then a roller mill. Since montmorillonite is well dispersed in water and to keep the gangue minerals in coarse sizes, no extra crushing and/or grinding process is applied.

2. After the size reduction, 20 g of crushed raw montmorillonite sample is added to a beaker which contains 2 L of distilled water.

3. The clay water dispersion obtained is kept for 24 h and then mixed with a mechanical mixer at 1000 rpm for 30 min.

4. After the mixing, the sedimentation process is started and the dispersion is kept for 8 h to complete settle the coarse gangue mineral particles. In this process, much finer particles of montmorillonite are suspended in water.

5. The dispersion on the sediment is siphoned (decantation) up to 3 cm height of the sediment as shown in **Figure 2**.

6. In order to prevent the settled gangue minerals from mixing with the suspended clay minerals as a result of the turbulence effect, the siphoning process is continued from top to bottom slowly up to a height of 3 cm from the bottom.

7. Finally, the montmorillonite particles in the dispersion obtained after siphoning are separated by centrifugation or filtration and then dried at 35-50°C for 24 h.

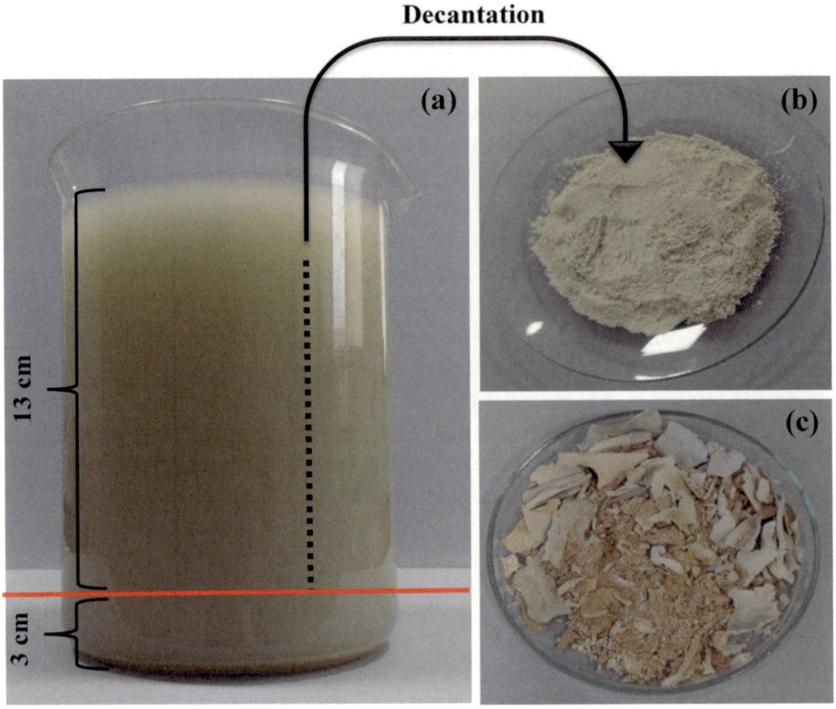


**Figure 2.**
*An example of sedimentation process; (a) raw montmorillonite dispersion, (b) purified montmorillonite, (c) settled gangue minerals [1].*

### 2.1.2 Centrifugal separation

One of the physical purification methods performed for separating montmorillonite clay and gangue minerals according to the difference in particle size is centrifugal separation. Centrifuge is a technique in which centrifugal force is used to separate various components in a liquid from each other or from the liquid. In solid–liquid separation processes, centrifugation is performed by rotating a closed container containing liquid and solid particles radially at high speed. It works by causing denser and/or larger particles to move outward in the radial direction, while less dense and/or smaller objects shift and move towards the center. In solid–solid separation processes, an aqueous media containing two or more different solids moves by following two different ways under the effect of centrifugal force. As the water moves in both directions, heavy and/or coarse particles move on one way and less dense and/or small particles moves on the other way. For this technique hydrocyclone, Falcon gravity concentrator (FGC), Knelson gravity concentrator (KGC), and multi gravity separator (MGS) devices are generally used (**Figure 3**).

Hydrocyclone (**Figure 3a**) is one of the most used mineral processing equipment in the plants that can be used to separate different minerals from each other based on both their particle size and specific weight. In some cases it is also used as dewatering equipment. In montmorillonite purification processes, multi stage hydrocyclone systems should be used to ensure maximum yield and purification degree (**Figure 4**). Multi gravity separator (MGS) (**Figure 3b**) are generally used to beneficiate minerals which liberation particle sizes are under 500 μm. MGS consists of an inclined drum that rotates around its own axis while shaken axially in a sinusoidal motion. Inside the drum is a scraper assembly that rotates in the same direction as the drum, but at a slightly faster speed. MGS also separates minerals based on particle size difference and therefore clay ores such as montmorillonite

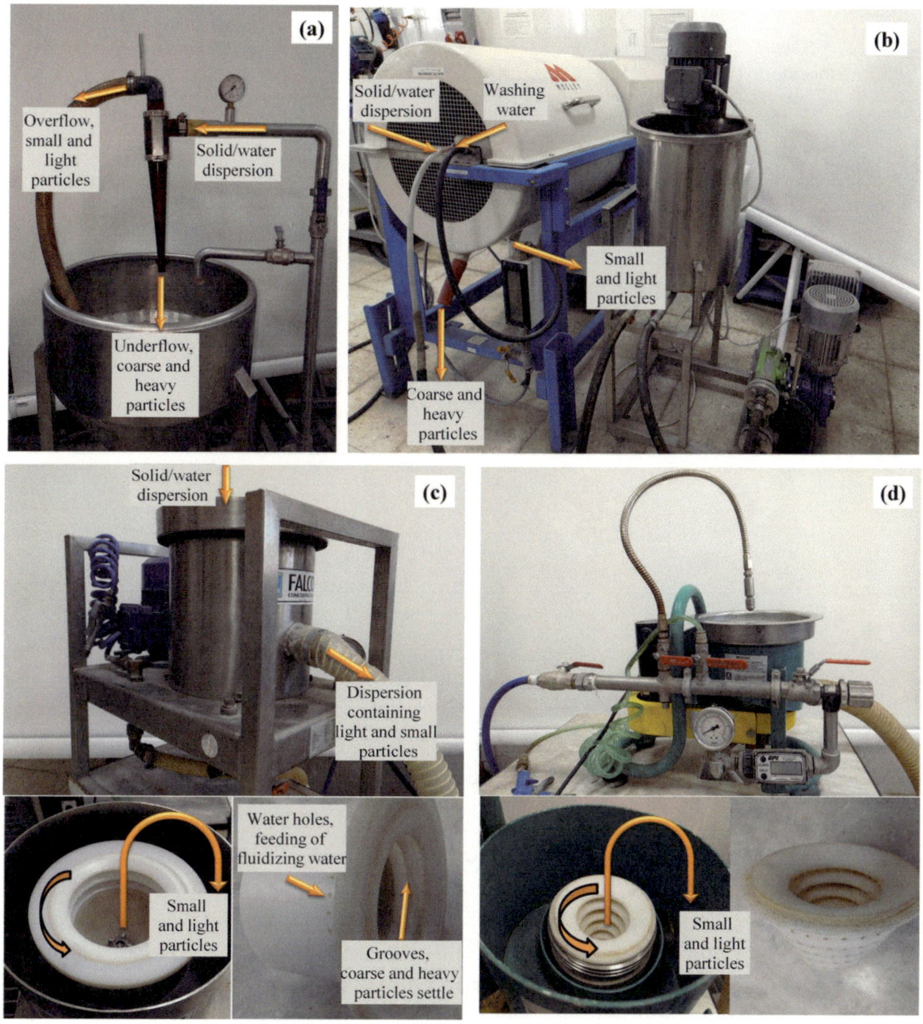


**Figure 3.**
*Photographs of laboratory scale (a) hydrocyclone, (b) multi gravity separator, (c) falcon gravity concentrator, (d) Knelson gravity concentrator.*

can be purified using MGS. FGC (**Figure 3c**) and KGC (**Figure 3d**) are the similar devices and work as batch system. There is a slight structural difference between them and the entire bowl in the KGC is grooved while a part of the Falcon bowl is grooved. Maximum 5 wt.% solid ratio should be used when using all centrifugal separators to ensure maximum yield and purification degree.

### *2.1.3 Aero separation*

Aero separation is based on the different motion characteristics of mineral particles in various devices working with air flow media. Aero classifiers (separators) are divided into two main groups as static and dynamic separators, and classify mineral particles according to their size, shape, and specific weight. In aero cyclone and static separators, the ore is fed to the classifier at a certain speed and classification is performed under the effect of this speed. In dynamic separators the motion of particles is controlled by both air and a blade stator rotating at a certain speed. Dynamic separators are the most used type among aero separators because of its high efficiency and capacity. The figures of the some aero separators are given in **Figure 5**.

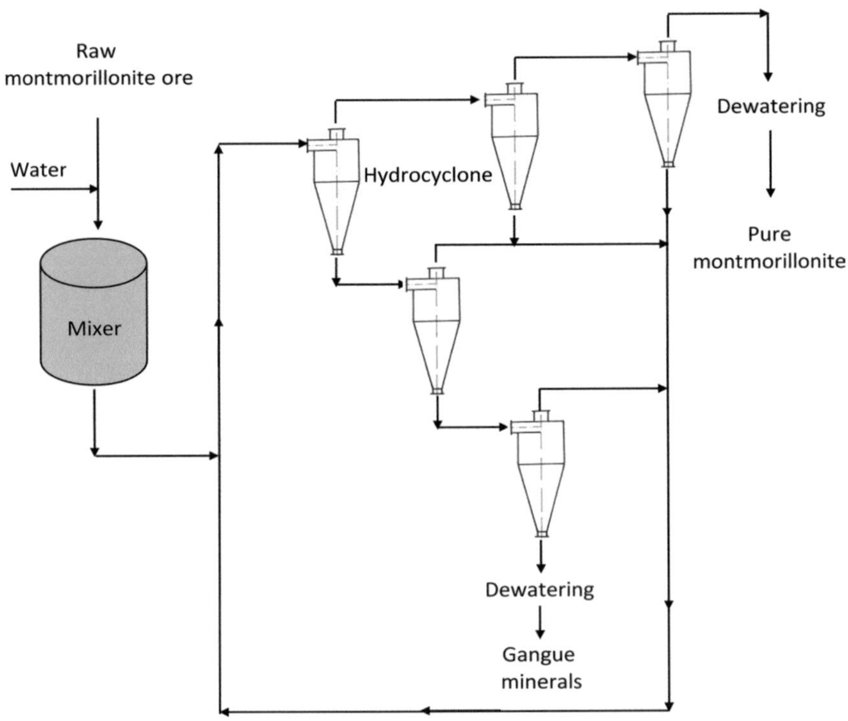


**Figure 4.**
*A simple flow sheet for montmorillonite purification using multi stage hydrocyclone process.*

The working principles of aero separators are similar. For the classification process, the ground ore is fed into the separator along with the air and suspended. The mineral particles are moved under the gravity and centrifugal forces according to their specific weight, shape, and size. The combination of these forces determines the movement direction of the particles, and the particles are separated from each other. The particles move with the air if the particle size and weight are small enough. The value at which gravity and centrifugal force are in equilibrium is called the separation size or "cut point". In other words, the cut point of a separator defines that the particles having a size of cut point value moves with both coarse and small particles.

Grinding is the most important process in the montmorillonite purification using aero separators. Because gangue minerals become thinner to the particle size of clay minerals at the end of extreme grinding and this is an undesirable situation. For an efficient purification, there should be a significant particle size difference between montmorillonite and gangue minerals. To ensure this, grinding should always be controlled by particle size measurements. Apart from this, different grinding media such as glass and/or plastic based balls can be used instead of steel balls. The specific gravity of glass and plastic balls is less than steel balls, and this ensures that the softer clay minerals are largely dispersed, while the harder gangue minerals are not grinded too much and remain in coarse sizes.

In montmorillonite purification processes using aero separators multi stage processes should be used to obtain maximum yield and purification degree. **Figure 6** shows an example of simple flow sheet for montmorillonite purification process using multi stage dynamic separator. The number of separators in the process can be increased or decreased by prior experimental studies.

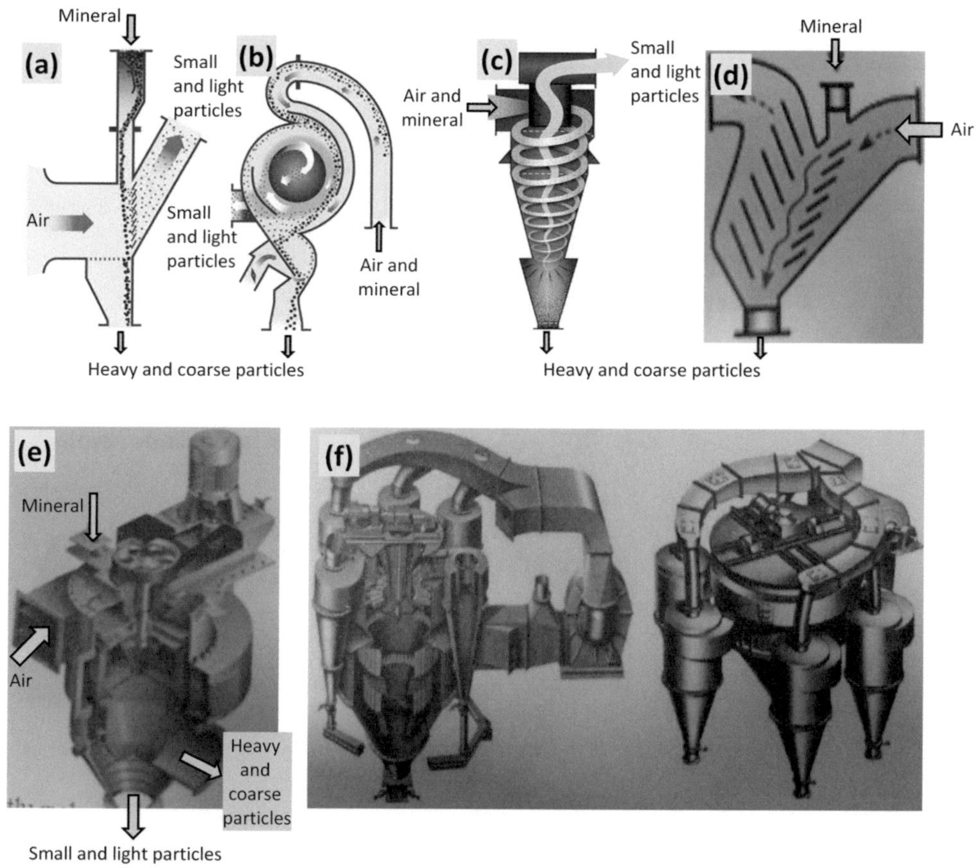


**Figure 5.**
*Aero separators, (a) gravitational static separator, (b) centrifugal static separator, (c) aero cyclone static separator, (d) v-type static separator, (e) dynamic separator, (f) dynamic and cyclonic separator [13, 14].*

### 2.1.4 Sieving

Sieving is the most used process part in mineral processing plants. Minerals are classified according to their particle sizes by various different sieves. Montmorillonite purification can be performed using both wet and dry sieving processes. Since the montmorillonite particles are very well dispersed in water (<2 μm), there is no need to pre-grind the raw ore in the wet sieving purification process. In this way, it is ensured that the gangue minerals remain in coarse sizes. However, if the dry sieving process is to be performed, it is necessary to do a controllable grinding before the sieving in order to disperse the agglomerated clay mineral particles. The yield and the purification degree of the montmorillonite purification process by sieving are in most cases lower than other methods because the sieving efficiency of the sieves in fine sizes decreases too much.

## 2.2 Chemical purification (leaching)

Leaching is the process of extracting organic or inorganic substances from a solid by dissolving them in a liquid. It is generally used for the extraction of valuable minerals (especially gold and rare earths) from raw ores, and the process is classified as chemical beneficiation (enrichment) method. However, in some cases, it can also be used to remove waste (gangue) minerals from valuable ores. Chemical clay

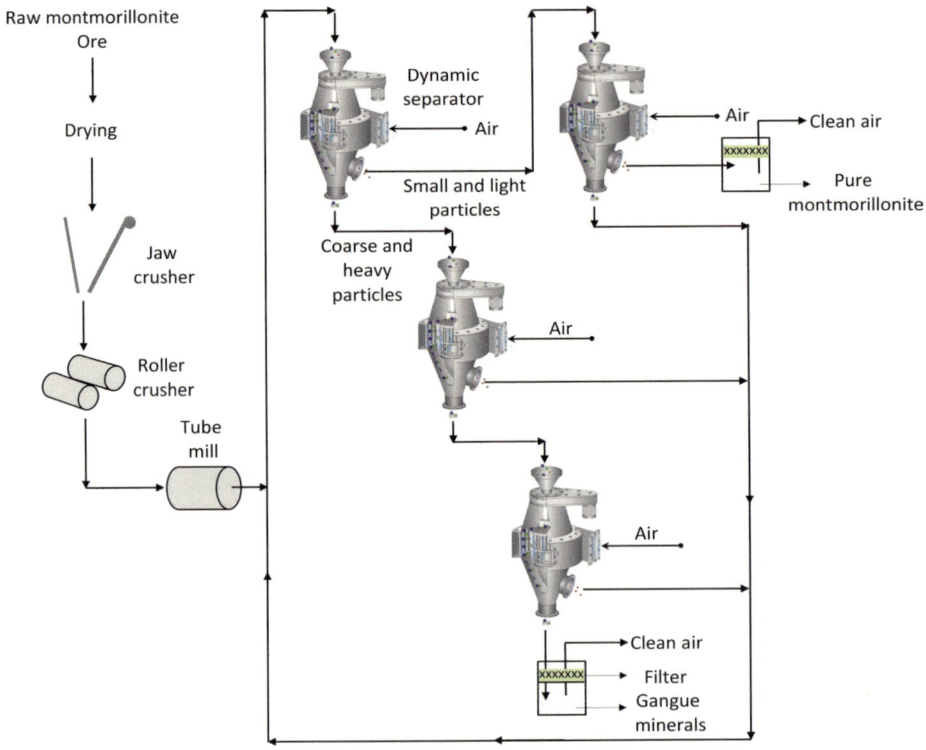


**Figure 6.**
*A typical flow sheet for montmorillonite purification using multi stage dynamic separator.*

purification is one of them and is based on decomposition of carbonates, dissolution of hydroxides, oxidation of organic materials, and dissolution of silica.

Carbonated minerals such as calcite dissolve easily in dilute hydrochloric (HCl) acid, and acetic acid ($CH_3COOH$). For this reason, raw montmorillonite clay containing carbonated minerals can be purified from carbonates by first treating with dilute HCl or $CH_3COOH$. During acid treatment, the pH value of the clay dispersion should be kept under constant control and the pH value should be prevented from falling below 4.5. Below this pH value, the clay structure can be damaged in a long time [15]. After acid treatment, the edges of the crystals are opened and the $Al^{3+}$ and $Mg^{2+}$ cations of the octahedral layers become soluble and at the same time the catalytic properties of the clay increase. Since an increase in the specific surface area of acid-treated clays is observed, the adsorption capacity of the clays also increases [16–18].

Iron, aluminum, and manganese (hydr)oxides can be removed from the clay ore by complexing multivalent cations with citrate. There are amorphous and crystal forms of free iron oxides in most of clay ores. $Fe^{3+}$ should be reduced to $Fe^{2+}$, which when treated with sodium dithionite ($Na_2S_2O_4$) forms a stable citrate complex. The use of citrate with sodium dithionite not only helps iron extraction, but also removes some alumina coatings, thus helping to dissolve free silica cements stabilized with alumina coatings [19, 20]. **Table 1** shows several chemicals which were tested for dissolution of iron hydroxides. For all chemical combinations, the same procedure outlined below can be performed. Each reaction given in **Table 1** is carried out with stirring for 3 hours and repeated twice. The final solid product obtained is centrifuged and washed at least twice with 0.5 mol/L sodium chloride (NaCl) solution. Finally, samples are washed several times in distilled water to remove residual NaCl [21].

| Method | Used chemicals |
|---|---|
| A | $Na_3C_6H_5O_7$ (sodium citrate, 29.4 g), NaOH (0.5 mol/L), $Na_2S_2O_4$ (sodium dithionite, 0.5 g), pH ~ 7 |
| B | $(NH_4)_2C_2O_4$ (ammonium oxalate, 0.25 mol/L), $H_2C_2O_4$ (oxalic acid, 0.1 mol/L), pH ~ 3 |
| C | $(NH_4)_2C_2O_4$ (ammonium oxalate, 0.25 mol/L), $Na_2S_2O_4$ sodium dithionite (0.5 g) |
| D | $HNO_3$ (0.5 mol/L) |

**Table 1.**
*Chemicals used for dissolution of iron hydroxides from montmorillonite [21].*

Another recommended procedure for removing of iron hydroxides can be summarized as follow. First, 15 g of montmorillonite ore is dispersed in 250 mL water and then prepared dispersion is added in a 200 mL buffer solution consisting of 0.3 M sodium citrate, 1 M sodium hydrogen carbonate, and 1.2 M HCl. The mixture is then heated to 75°C and 4 g sodium dithionite is added. After 30 min mixing, solid liquid separation is applied and then the solid material is washed with 200 mL 0.05 M HCl. After washing, the reduction process is repeated and then the washing is performed with a mixture of 200 mL 0.5 M NaCl and 200 mL 0.5 M sodium acetate. Finally, the resulting solid product is added to 1000 mL solution (250 mL 30% $H_2O_2$ + 750 mL 0.5 M sodium acetate) and kept at 70°C for 30 min to ensure oxidation of the organic substances. Following the oxidation process, solid–liquid separation is performed and organic substances are removed by washing with 200 mL 0.5 M NaCl twice [15, 22].

Organic substances can be removed from clay ores by oxidizing using hydrogen peroxide ($H_2O_2$), sodium hypochlorite (NaOCl), aqueous bromine (Br), and sodium peroxodisulphate ($N_2S_2O_8$) in some buffer solutions such as sodium hydrogen carbonate and sodium tetraborate [23–28]. By using $Na_4P_2O_7$ as a dispersing agent in the $H_2O_2$ treatment, the amount of organic carbon removal can be increased [29]. Kaiser and Guggenberger [30] investigated a modified NaOCl treatment on many soil samples and found that from 77–95% of the initial organic carbon could be removed from soil. It has been suggested that the efficiency of removing organic matter by $Na_2S_2O_8$ process is superior to $H_2O_2$ and NaOCl [31].

The amorphous silica in the clay ore can be removed easily by dissolving in a boiling solution containing 5%wt. sodium carbonate ($Na_2CO_3$). Since this reaction makes the leaching of hydroxides difficult, it should be used when all the chemical processes described above are completed [15].

## 3. Evaluation of the purification methods

The efficiency of each purification technique can be assessed based on XRD, XRF, particle size, cation exchange capacity (CEC), swelling index (SI) analyzes before and after experiments. The montmorillonite content can be calculated using quantitative XRD data and rietveld analysis. The yield is obtained by calculating the percentage by weight of the product obtained after beneficiation to the fed product. Apart from these, recovery can be calculated from Eq. (2) [32].

$$R(\%) = \frac{M_c * CEC_c}{M_f * CEC_f} \tag{2}$$

Here; R is the recovery (%), $M_c$ and $M_f$ are the mass of concentrate and feed (%), $CEC_c$ and $CEC_f$ are cation exchange capacities of concentrate and feed, respectively.

For an example, **Figure 7** shows a sample of XRD patterns of two different type montmorillonite ores before and after purification by sedimentation and centrifugal separation techniques. The sedimentation process was performed exactly as described in the Section 2.1.1. FGC was used for centrifugal separation and the process used can be summarized as follow. Beneficiation trials were performed at 300 G centrifugal force and 1.5 L/min suspension feed rate. For this, 1.0% clay dispersion was prepared as described in the sedimentation process and then left for 10 minutes to allow the coarse particles to settle. The dispersion remaining on the collapsed particles was completely removed and fed to FGC once [1].

According to **Figure 7**, most of the gangue minerals were removed from raw montmorillonite-a by sedimentation and FGC. However, some of quartz minerals and little amounts of kaolinite remained in montmorillonite-b after both sedimentation and FGC processes. This can be explained by the fact that the quartz and kaolinite particles smaller than 5 μm are suspended in water and moves with montmorillonite particles. Small amounts (<3%) of such small particles that could not be detected by the XRD device may also have remained in the montmorillonite-a products. Considering the degree of purification, XRD data did not show any significant difference between sedimentation and FGC techniques for both clays. Apart from XRD analysis, cation exchange capacity (CEC), swelling index (SI), and yield were also calculated before and after purification processes. The results are summarized in the **Table 2** [1].

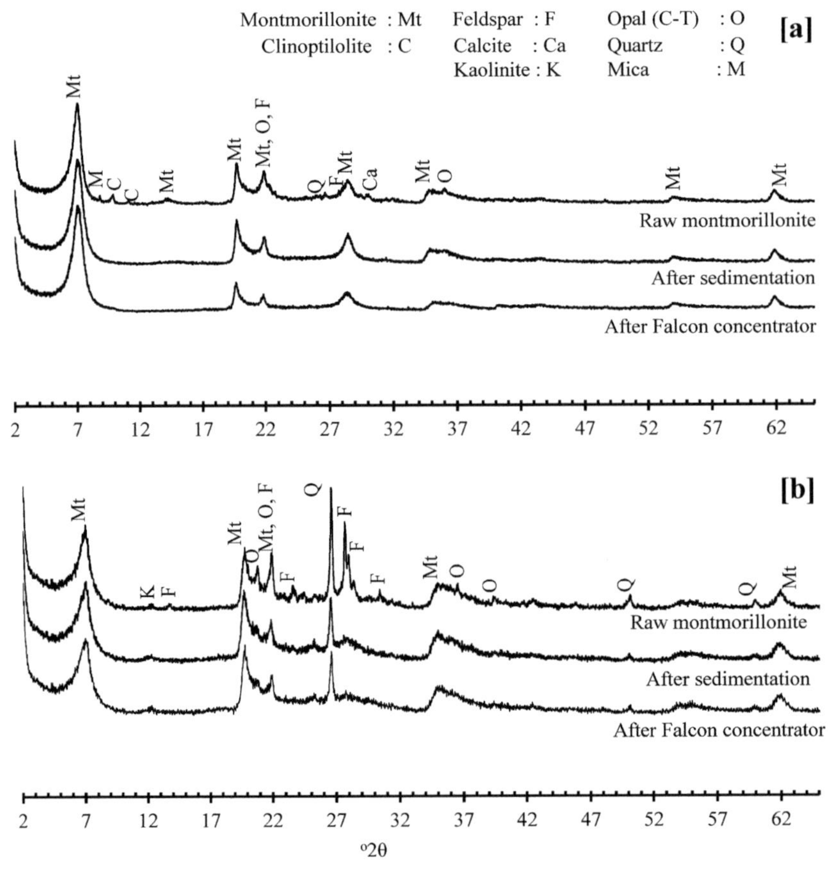


**Figure 7.**
*XRD patterns of two different type montmorillonite clays before and after beneficiation processes, (a) montmorillonite-a, (b) montmorillonite-b [1].*

| Sample | CEC (meq/100 g) | SI (mL) | Yield (%) |
|---|---|---|---|
| Montmorillonite-a (before purification) | 80 | 25 | — |
| Montmorillonite-a (after sedimentation) | 95 | 27 | 67 |
| Montmorillonite-a (after FGC) | 92.5 | 27 | 81 |
| Montmorillonite-b (before purification) | 60 | 9 | — |
| Montmorillonite-b (after sedimentation) | 90 | 11 | 45 |
| Montmorillonite-b (after FGC) | 87.5 | 11 | 68 |

**Table 2.**
*CEC, SI, and yields of montmorillonite samples before and after purification.*

Although the montmorillonite contents in purified products were almost the same according to XRD patterns and the data in **Table 2**, much higher yields were obtained by FGC compared to the sedimentation technique. Therefore, it would be a better choice to use FGC to beneficiate montmorillonite clays in a shorter time and with higher yields.

In another study which investigated the purification and modification of montmorillonite reported a combination of sedimentation, modification, flotation and FGC techniques. According to reported results, a product of purified and organically modified montmorillonite was obtained with 75% yielding, and 97% montmorillonite content calculated from quantitative XRD analysis [33].

In a study, purification of montmorillonite was investigated by centrifugal separators and it was reported that commercial centrifuge equipment such as Multi Gravity separator, Hydrocyclone, FGC, and KGC operating with a residence time of 10–15 seconds and G forces less than 300 G are not suitable for obtaining pure quality montmorillnoite. This is the case for montmorillonite ores containing very small sizes (<10 μm) gangue minerals. In addition, FGCs can be operated up to 300 G forces and 1–2 inch Mozley type hydrocyclones can be used up to 3500 G forces at 7–8 bar feed pressures [34].

Particle size distributions before and after beneficiation can also be evaluated to roughly see the degree of purification. **Figure 8** reveals that there are particles larger than 10 μm in raw clay sample which can be attributed to gangue minerals. As expected, all purified clay samples shows narrower size distributions because of removed coarse gangue minerals [35]. Another research group was also reported close results on this issue and reported that the particle size distribution of montmorillonite purified by sedimentation is between 0.3–7.5 μm [36].

Hydrocyclone is also most used beneficiation equipment for montmorillonite [37–39]. It has some advantages such as continuous system, high capacity, and working with higher solid ratio when compared to sedimentation, FGC, and KGC techniques. Multi stage hydrocyclone system should be used to ensure sufficient yield ratios (approximately 75%), however. The solid ratio is one of the most important parameter that should be around 5%. When higher solid ratios are used, separation is adversely affected due to problems such as particle-particle interactions and obstruction of the apex outlet due to high swelling of montmorillonite.

In a detailed study of multi-stage hydrocyclone application, two hydrocyclones with a large vortex finder diameter (14.3 mm) and then a smaller one (8.0 mm) were run in series with each other. As a result, purified montmorillonite with a cation exchange capacity (CEC) of 98 meq/100 g and 75% yield, which has the properties required for commercial grade drilling mud, water/solvent based bentonite, and sub material for nanocomposite technology, was obtained. In addition, waste with a CEC of 27–32 meq/100 g and 15% yield was obtained. The CEC of montmorillonite sample before beneficiation process was calculated to be 78 meq/100 g [39].

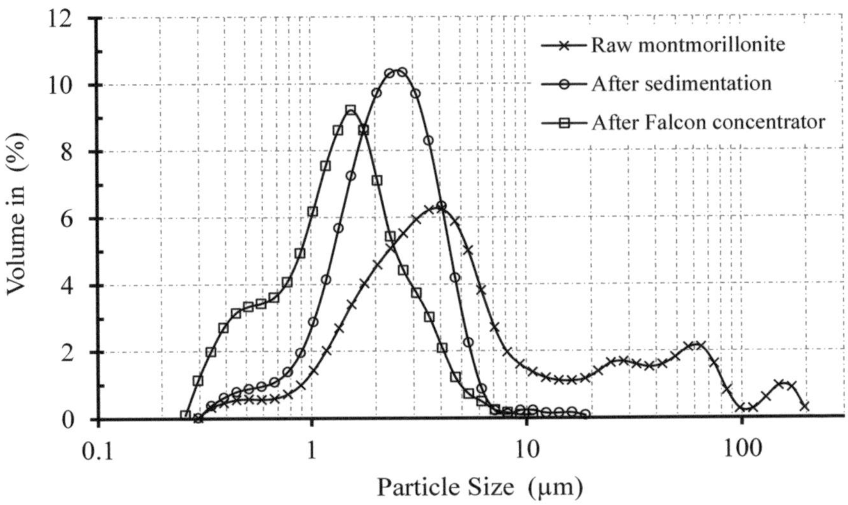


**Figure 8.**
*An example of particle size distributions of a montmorillonite clay sample before and after beneficiation processes [35].*

Conventional decanter centrifuge devices apart from FGC, KGC, MGS, and hydrocyclones were also used to beneficiate montmorillonite ores. For an example, Gong et al. [8] investigated purification of montmorillonite using conventional centrifuge device. The process performed in this study combines grinding, dispersion, and centrifugal separation, respectively. Maximum 96.5% montmorillonite content with 71% yield was obtained by optimal parameters (grinding time 60 min, dispersant dosage 1%, centrifuge speed 700 rpm, and centrifuge time 2 min). Although products with high montmorillonite content and efficiency are obtained, the capacity of decanter centrifuges is very low compared to other techniques and therefore their usage areas on an industrial scale are limited. The centrifugation time for decanter centrifuge devices can be calculated using the following equation based on Stockes' law.

$$t = \left[\eta \log_{10}(R/S)\right] / \left[3.81 N^2 r^2 \Delta S\right] \tag{3}$$

Here; t: centrifugation time (second), η: viscosity of the fluid (poise), S: distance from the suspension surface to the ax of rotor (cm), R: distance from the deposit surface to the ax of rotor (cm), N: rotation speed (rpm), r: maximum radius of the desired particles (cm), ΔS: specific gravity difference between the particles and the liquid suspension ($gcm^{-3}$).

Appropriate dispersion of clay minerals before purification is one of the most important stage of physical purification processes. The more successful the dispersion process, the more successful the purification rate and yield. Sodium hexametaphosphate ($Na(PO_3)_6$) is generally used as a dispersant agent for clay minerals. $Na(PO_3)_6$ is used at grinding or dispersion stages and is generally used in the range of 0.5–3% according to the clay weight. Zeta potential analysis can be used to determine exact amount of dispersant. In the measurements performed against the amount of dispersant, the amount of dispersant at the point where the zeta potential is fixed can be taken as the optimum value.

Most of the gangue minerals in the raw montmorillonite ores can be removed by physical purification methods. However, the particle sizes of gangue minerals are very important here. Some montmorillonite ores may contain very fine (<5 μm) gangue minerals. Such small particles occasionally settle between the clay layers and/or form small aggregates with clay particles. Consequently, the large particles settle in the dispersion, while the suspended fine gangue particles move with the clay minerals in aqueous or air media and their separation is not possible. In this case, chemical purification methods can be preferred either alone or in combination with physical methods in cases where products with high montmorillonite content (>%98) and yields are desired.

In a study chemical purification methods which listed in **Table 1** were used. The efficiency of the purification methods was assessed by elementary analysis and X-ray fluorescence analysis. C and D methods showed best results for two different montmorillonite ore that iron hydroxide amounts decreased from 11.34% to 2.21% and from 3.88% to 0.33%, respectively. As a result, it is reported that the (hydr) oxides were completely removed, and very small amount of remaining iron belongs to the montmorillonite structure. In addition, the clay structure was not affected severely at the end of leaching trials [21].

Menegatti et al. [24] investigated removal of organic matter from clay ores using disodium peroxodisulphate ($N_2S_2O_8$). Sodium hydrogen carbonate ($NaHCO_3$) was used as buffer. It is reported that the use of $NaHCO_3$ as a buffer instead of $Na_2HPO_4$ and $Na_2B_4O_7$ was more efficient. The leaching technique was performed at 80°C for 60 min and the buffer/oxidant mass ratio was kept as 1:1. As a result, more than 98% of organic substances were removed from clay ores. The authors also stated that their new method was more effective than conventional oxidative procedures using hydrogen peroxide, sodium hypochlorite, and ozone.

## 4. Conclusion

Purification of montmorillonite was investigated in two main classes as physical and chemical purification. There are generally physical purification methods in the literature because of simple and fast application process. However, in physical methods, the purification yield and purification degree (montmorillonite content) generally remain between 65 and 80% and 70–95%, respectively. This depends on the type of ore, method, and process used. Sedimentation, hydrocyclone, and Falcon gravity concentrator give better results compared to other physical enrichment methods working in aqueous environment when used at optimal parameters mentioned above. Dynamic aero separators can also show good results and they are good candidates for dry clay beneficiation methods. However, there is no literature knowledge that dynamic separators are used for clay purification. A detailed research study on clay beneficiation using dynamic separators will be a good study to show the clay enrichment efficiency with dry methods.

In physical purification methods multi stage processes for hydrocyclone, centrifugal separator, and aero separator increase the yield and purification degree. Some montmorillonite ores contain very thin (<5 μm) quartz and other mineral particles. These particles cannot be separated easily using sedimentation, centrifugal separators, and sieving techniques.

In this context, leaching (chemical purification) is useful technique for montmorillonite purification in cases highest yield (>90%) and montmorillonite content (>%95) is desired. In the chemical purification process, leaching operations should be performed in the following order; dissolution of carbonate minerals, dissolution of metal hydroxides, oxidation of organic substances, and finally silica dissolution. After each stage, the washing process should be done at a sufficient level.

## Author details

Hakan Ciftci
Department of Mining, Faculty of Engineering, Afyon Kocatepe University, Afyonkarahisar, Turkey

*Address all correspondence to: hakanciftci86@gmail.com

## References

[1] Çiftçi H, Ersoy B, Evcin A. Purification of Turkish Bentonites and Investigation of the Contact Angle, Surface Free Energy and Zeta Potential Profiles of Organo-Bentonites as a Function of CTAB Concentration. Clays and Clay Minerals. 2020;68(3):250-261. DOI: 10.1007/s42860-020-00070-0

[2] Yang JH, Lee JH, Ryu HJ, Elzatahry AA, Alothman ZA, Choy JH. Drug–clay nanohybrids as sustained delivery systems. Applied Clay Science. 2016;130:20-32. DOI: 10.1016/j.clay.2016.01.021

[3] Nones J, Riella HG, Trentin AG, Nones J. Effects of bentonite on different cell types: A brief review. Applied Clay Science. 2015;105-106: 225-230. DOI: 10.1016/j.clay.2014.12.036

[4] Jain S, Datta M. Oral extended release of dexamethasone: Montmorillonite–PLGA nanocomposites as a delivery vehicle. Applied Clay Science. 2015;104:182-188. DOI: 10.1016/j.clay.2014.11.028

[5] Jayrajsinh S, Shankar G, Pharm M, Agrawal YK, Bakre L. Montmorillonite nanoclay as a multifaceted drug-delivery carrier: A review. Journal of Drug Delivery Science and Technology. 2017;39:200-209. DOI: 10.1016/j.jddst.2017.03.023

[6] Elmore SE, Mitchell N, Mays T, Brown K, Marroquin-Cardona A, Romoser A, Phillips TD. Common African cooking processes do not affect the aflatoxin binding efficacy of refined calcium montmorillonite clay. Food control. 2014;37:27-32. DOI: 10.1016/j.foodcont.2013.08.037

[7] Summa V, Tateto F. Clay minerals as adsorbents of aflatoxin M1 from contaminated milk and effects on milk quality. Applied Clay Science. 2014;88: 92-99. DOI: 10.1016/j.clay.2013.11.028

[8] Gong Z, Liao L, Lv G, Wang X. A simple method for physical purification of bentonite. Applied Clay Science. 2016;119:294-300. DOI: 10.1016/j.clay.2015.10.031

[9] Shah LA, Valenzuela MGS, Ehsan AM, Díaz FRV, Khattak NS. Characterization of Pakistani purified bentonite suitable for possible pharmaceutical application. Applied Clay Science. 2013;83(84):50-55. DOI: 10.1016/j.clay.2013.08.007

[10] Çiftçi H, Arpa MD, Gülaçar İM, Özcan L, Ersoy B. Development and evaluation of mesoporous montmorillonite/magnetite nanocomposites loaded with 5-Fluorouracil. Microporous and Mesoporous Materials. 2020;303: 110253. DOI: 10.1016/j.micromeso.2020.110253

[11] Bergaya F, Lagaly G. General introduction: clays, clay minerals, and clay science. In: Bergaya F, Theng BKG, Lagaly G, editors. Handbook of Clay Science. Amsterdam: Elsevier; 2006. p. 1-18. DOI: 10.1016/S1572-4352(05)01001-9

[12] Lagaly G, Dekany I. Colloid clay science. In: Bergaya F, Theng BKG, Lagaly G, editors. Handbook of Clay Science. Amsterdam: Elsevier; 2013. p. 243-345. DOI: 10.1016/B978-0-08-098258-8.00010-9

[13] Metso Corporation. Basics in Mineral Processing Handbook. 10th ed. Helsinki: Metso Corporation; 2015. 354 p.

[14] Yıldız N. Cevher Hazırlama ve Zenginleştirme. 1th vol. Ankara: Ertem Basım; 2014. 730 p.

[15] Carrado KA, Decarreau A, Petit S, Bergaya F, Lagaly G. Synthetic clay minerals and purification of natural

clays. In: Bergaya F, Theng BKG, Lagaly G, editors. Developments in Clay Science 1, Handbook of Clay Science. Vol. 1, UK; Elsevier; 2006. p. 115-139. DOI: 10.1016/S1572-4352(05)01004-4

[16] D'Amico DA, Ollier RP, Alvarez VA, Schroeder WF, Cyras VP. Modification of bentonite by combination of reactions of acid-activation, silylation and ionic exchange. Applied Clay Science. 2014;99:254-260. DOI: 10.1016/j.clay.2014.07.002

[17] Hou D, Hu S, Huang Y, Gui R, Zhang L, Tao Q, Zhanga C, Tian S, Komarneni S, Ping Q. Preparation and in vitro study of lipid nanoparticles encapsulating drug loaded montmorillonite for ocular delivery. Applied Clay Science. 2016;119:277-283. DOI: 10.1016/j.clay.2015.10.028

[18] Zhu R, Chen Q, Zhou Q, Xi Y, Zhu J, He H. Adsorbents based on montmorillonite for contaminant removal from water: A review. Applied Clay Science. 2016;123:239-258. DOI: 10.1016/j.clay.2015.12.024

[19] Mehra OP, Jackson ML. Iron oxide removal from soils and clays by a dithionitecitrate system buffered with sodium bicarbonate. Clays and Clay Minerals. 1960;7:317-327. DOI: 10.1346/CCMN.1958.0070122

[20] Holmgren GGS. A rapid citrate–dithionite extractable iron procedure. Soil Science Society of America Proceedings. 1967;31:210-211. DOI: 10.2136/sssaj1967.03615995003100020020x

[21] Thuc CNH, Grillet AC, Reinert L, Ohashi F, Thuc HH, Duclaux L. Separation and purification of montmorillonite and polyethylene oxide modified montmorillonite from Vietnamese bentonites. Applied Clay Science. 2010;49:229-238. DOI: 10.1016/j.clay.2010.05.011

[22] Stul MS, van-Leemput L. Particle-size distribution, cation exchange capacity and charge density of deferrated montmorillonites. Clay Minerals. 1982;17:209-215. DOI: 10.1180/claymin.1982.017.2.06

[23] Mikutta R, Kleber M, Kaiser K, Jahn R. Review: Organic matter removal from soils using hydrogen peroxide, sodium hypochlorite, and disodium peroxodisulfate. Soil Science Society of America Journal. 2005;69:120-135. DOI: 10.2136/sssaj2005.0120

[24] Menegatti AP, Früh-Green GL, Stille P. Removal of organic matter by disodium peroxodisulphate: effects on mineral structure, chemical composition, and physicochemical properties of some clay minerals. Clay Minerals. 1999;34:247-257. DOI: 10.1180/000985599546217

[25] Mitchell BD, Smith BFL. The removal of organic matter from soil extracts by bromine oxidation. Journal of Soil Science 1974;25:239-241. DOI: 10.1111/j.1365-2389.1974.tb01120.x

[26] Anderson JU. An improved treatment for mineralogical analysis of samples containing organic matter. Clays and Clay Minerals. 1963;10:380-388. DOI: 10.1346/CCMN.1961.0100134

[27] Robinson WO. The determination of organic matter in soils by hydrogen peroxide. Journal of Agricultural Research. 1927;34:339-356.

[28] Mikutta R, Kleber M, Kaiser K, ET AL. Review: organic matter removal from soils using hydrogen peroxide, sodium hypochlorite, and disodium peroxodisulfate. Soil Science Society of America Journal. 2005;69(1):120-135. DOI: 10.2136/sssaj2005.0120

[29] Simon NS, Hatcher SA, Demas S. Comparison of methods for removal of organic carbon and extraction of

chromium, iron and manganese from an estuarine sediment standard and sediment from the Calcasieu river estuary, Louisiana, USA. Chememical Geology. 1992;100:175-189. DOI: 10.1016/0009-2541(92)90111-H

[30] Kaiser K, Guggenberger G. Mineral surfaces and soil organic matter. European Journal of Soil Science. 2003;54:1-18. DOI: 10.1046/j.1365-2389.2003.00544.x

[31] Meier LP, Menegatti AP. A new, efficient, one-step method for the removal of organic matter from clay-containing sediments. Clay Minerals. 1997;32:557-563. DOI: 10.1180/claymin.1997.032.4.06

[32] Grim, RE. Clay Mineralogy, 2nd Ed. New York: McGraw-Hill; 1968. 596 p.

[33] Yeşilyurt Z, Boylu F, Çinku K, Esenli F, Çelik MS. Simultaneous purification and modification process for organobentonite production. Applied Clay Science. 2014;95:176-181. DOI:10.1016/j.clay.2014.04.008

[34] Boylu F, Hojiyev R, Ersever G, Ulcay Y, Çelik MS. Production of Ultrapure Bentonite Clays through Centrifugation Techniques. Separation Science and Technology. 2014;47(6):842-849. DOI: 10.1080/01496395.2011.634475

[35] Çiftçi H. Silica Pillared Magnetic Clay Nanocomposites and Drug Release. Vol. 1, Moldova: Lambert Academic Publishing; 2020. 160 p.

[36] Shah LA, Khattak NS, Valenzuela MGS, Manan A, Valenzuela Diaz FR. Preparation and characterization of purified Na-activated bentonite from Karak (Pakistan) for pharmaceutical use. Clay Minerals. 2013;48:595-603. DOI: 10.1180/claymin.2013.048.4.03

[37] de Oliveira PL, de Figueirêdo JMR, Cartaxo JM, Neves GA, Ferreira HC. Influence of Hydrocycloning Pressure on the Purification of Bentonite. Materials Science Forum. 2014;798-799:55-60. DOI:10.4028/www.scientific.net/msf.798-799.55

[38] Gama AJA, Neves GA,Barros PL, Neto ATP, Alves JJN. Hydrocyclone performance for bentonite clay purification.Chemical Engineering Research and Design. 2020;161:168-177. DOI:10.1016/j.cherd.2020.07.005

[39] Boylu F, Çinku K, Esenli F, Çelik MS. The separation efficiency of Na-bentonite by hydrocyclone and characterization of hydrocyclone products. International Journal of Mineral Processing. 2010;94:196-202. DOI: 10.1016/j.minpro.2009.12.004